T0185715

SpringerBriefs in Electrical and Computer Engineering

Series Editors

Woon-Seng Gan, School of Electrical and Electronic Engineering, Nanyang Technological University, Singapore, Singapore

C.-C. Jay Kuo, University of Southern California, Los Angeles, CA, USA

Thomas Fang Zheng, Research Institute of Information Technology, Tsinghua University, Beijing, China

Mauro Barni, Department of Information Engineering and Mathematics, University of Siena, Siena, Italy

SpringerBriefs present concise summaries of cutting-edge research and practical applications across a wide spectrum of fields. Featuring compact volumes of 50 to 125 pages, the series covers a range of content from professional to academic. Typical topics might include: timely report of state-of-the art analytical techniques, a bridge between new research results, as published in journal articles, and a contextual literature review, a snapshot of a hot or emerging topic, an in-depth case study or clinical example and a presentation of core concepts that students must understand in order to make independent contributions.

More information about this series at http://www.springer.com/series/10059

Christoph Guger · Brendan Z. Allison ·
Kai Miller
Editors

Brain–Computer Interface Research

A State-of-the-Art Summary 8

 Springer

Editors
Christoph Guger
g.tec medical engineering GmbH
Schiedlberg, Austria

Kai Miller
Mayo Clinic
Rochester, MN, USA

Brendan Z. Allison
Department of Cognitive Science
University of California at San Diego
La Jolla, USA

ISSN 2191-8112   ISSN 2191-8120   (electronic)
SpringerBriefs in Electrical and Computer Engineering
ISBN 978-3-030-49582-4   ISBN 978-3-030-49583-1   (eBook)
https://doi.org/10.1007/978-3-030-49583-1

© The Editor(s) (if applicable) and The Author(s), under exclusive license to Springer Nature
Switzerland AG 2020
This work is subject to copyright. All rights are solely and exclusively licensed by the Publisher, whether
the whole or part of the material is concerned, specifically the rights of translation, reprinting, reuse of
illustrations, recitation, broadcasting, reproduction on microfilms or in any other physical way, and
transmission or information storage and retrieval, electronic adaptation, computer software, or by similar
or dissimilar methodology now known or hereafter developed.
The use of general descriptive names, registered names, trademarks, service marks, etc. in this
publication does not imply, even in the absence of a specific statement, that such names are exempt from
the relevant protective laws and regulations and therefore free for general use.
The publisher, the authors and the editors are safe to assume that the advice and information in this
book are believed to be true and accurate at the date of publication. Neither the publisher nor the
authors or the editors give a warranty, expressed or implied, with respect to the material contained
herein or for any errors or omissions that may have been made. The publisher remains neutral with regard
to jurisdictional claims in published maps and institutional affiliations.

This Springer imprint is published by the registered company Springer Nature Switzerland AG
The registered company address is: Gewerbestrasse 11, 6330 Cham, Switzerland

Contents

Introduction 1
Christoph Guger, Kai Miller, and Brendan Z. Allison

Generating Handwriting from Multichannel Electromyographic
Activity 11
Mikhail A. Lebedev, Alexei E. Ossadtchi, Elizaveta Okorokova,
Joseph S. Erlichman, Valery I. Rupasov, and Michael Linderman

Neural Decoding of Upper Limb Movements Using
Electroencephalography 25
Dingyi Pei, Martin Burns, Rajarathnam Chandramouli,
and Ramana Vinjamuri

Restoring Functional Reach-to-Grasp in a Person with Chronic
Tetraplegia Using Implanted Functional Electrical Stimulation
and Intracortical Brain-Computer Interfaces 35
A. Bolu Ajiboye, Leigh R. Hochberg, and Robert F. Kirsch

Towards Speech Synthesis from Intracranial Signals 47
Christian Herff, Lorenz Diener, Emily Mugler, Marc Slutzky,
Dean Krusienski, and Tanja Schultz

Neural Decoding of Attentional Selection in Multi-speaker
Environments Without Access to Clean Sources 55
James O'Sullivan, Zhuo Chen, Jose Herrero, Sameer A. Sheth,
Guy McKhann, Ashesh D. Mehta, and Nima Mesgarani

Goal-Directed BCI Feedback Using Cortical Microstimulation 65
Yohannes Ghenbot, Xilin Liu, Han Hao, Cole Rinehart, Sam DeLuccia,
Solymar Torres Maldonado, Gregory Boyek, Milin Zhang,
Firooz Aflatouni, Jan Van der Spiegel, Timothy H. Lucas,
and Andrew G. Richardson

**Neuromotor Recovery Based on BCI, FES, Virtual Reality
and Augmented Feedback for Upper Limbs**   75
Robert Gabriel Lupu, Florina Ungureanu, Oana Ferche,
and Alin Moldoveanu

**A Dynamic Window SSVEP-Based Brain-Computer Interface System
Using a Spatio-temporal Equalizer**   87
Chen Yang, Xiang Li, Nanlin Shi, Yijun Wang, and Xiaorong Gao

Highlights and Interviews with Winners   107
Christoph Guger, Brendan Z. Allison, and Kai Miller

Introduction

Christoph Guger, Kai Miller, and Brendan Z. Allison

Abstract Facebook, Elon Musk, and others are pursuing large-scale research and development projects involving brain-computer interface (BCI) systems. Each year, we have organized a BCI Research Award devoted to the best projects in BCI research. This year's book begins with a chapter that introduces BCIs, the Research Awards, and this book series. Most chapters present projects from the Eighth Annual BCI Research Award written by the scientists, doctors, and engineers behind each project. This book concludes with a chapter containing interviews with the winners, highlights from the BCI Research Awards Ceremony at the Seventh International BCI Meeting, and discussion of emerging BCI activities such as BCI Hackathons and Cybathlons.

Welcome to our eighth book! As we write or edit these chapters, we are currently preparing for the Tenth Annual Brain-Computer Interface (BCI) Research Award ceremony. Many of our earlier books have said the same thing that we often say at the awards ceremonies: we want to recognize and encourage top projects in the field. Each year, we still receive numerous award submissions, and the projects that have been nominated and their corresponding book chapters have been very interesting. We are grateful to readers like you for helping our books succeed, and hope you enjoy this year's book.

C. Guger (✉)
g.tec medical engineering GmbH, Schiedlberg, Austria
e-mail: guger@gtec.at

B. Z. Allison
Department of Cognitive Science, University of California at San Diego, La Jolla 92093, USA
e-mail: ballison@ucsd.edu

K. Miller
Mayo Clinic, Rochester, MN, USA
e-mail: kjmiller@gmail.com

© The Author(s), under exclusive license to Springer Nature Switzerland AG 2020
C. Guger et al. (eds.), *Brain–Computer Interface Research*,
SpringerBriefs in Electrical and Computer Engineering,
https://doi.org/10.1007/978-3-030-49583-1_1

1

1   What Is a BCI?

Brain-computer interfaces (BCIs) translate information from the brain directly to output signals like messages or commands. While most authors (including us) do not typically use a hyphen after "brain," we included it here to emphasize that BCIs read information from the brain, not write to it. BCIs provide a way to communicate without moving, and have primarily sought to help people with severe disabilities who cannot communicate any other way.

However, in the past several years, new BCI approaches for new types of patients have been developed and often validated with patients. The chapters in this book include BCIs that could potentially help patients hear more effectively, relearn how to move after a stroke, or use devices like exoskeletons or prostheses based on imagined movements. Another chapter from Bolu Ajiboye and colleagues shows a different BCI for prosthetic control using electrodes inside the skull, while Christian Herff and his team wrote a chapter about work toward BCIs that directly translate imagined speech into words.

Along with the many people working on BCIs, more people seem to be learning about BCIs. Universities offer new classes that include BCI research, with more faculty and other staff who can teach courses, mentor students, and support hands-on BCI activities and projects. Major public activities like BCI hackathons, Cybathlons, and demonstrations are also increasing.

2   The Annual BCI Research Award

The Annual BCI Research Award is organized through the non-profit BCI Award Foundation. Jury members may not submit projects, but otherwise, the award is open to any research group, regardless of their location, equipment used, etc. The awards procedure this year followed a procedure similar to prior years:

- A Chairperson of the Jury is chosen from a top BCI research institute.
- This Chairperson forms a jury of well-known BCI experts to judge the Award submissions.
- We post the submission instructions, scoring criteria, and the deadline for the Award.
- The jury scores each submission based on the scoring criteria below. The jury then chooses twelve nominees and the first, second, and third place winners.
- The nominees are announced online, asked to contribute a chapter to this annual book series, and invited to a Award Ceremony that is attached to a major conference (such as an International BCI Meeting or Conference).
- The Gala Awards Ceremony is a major part of each conference. Each of the twelve nominees are announced and come onstage to receive a certificate. Next, the winners are announced. The winner earns $3000 USD and the prestigious trophy. The 2nd and 3rd place winners get $2000 USD and $1000 USD, respectively.

This year, for the first time, the third-place prize was generously donated by the BCI Society. The BCI Society is a non-profit organization that organizes the BCI Meeting series (bcisociety.org), and has kindly made our Gala Awards Ceremony part of their meetings. The other cash prizes were provided by an Austrian company called g.tec medical engineering (author CG is the CEO), which manufactures equipment and software for BCIs and other applications.

The 2018 jury was:

Kai Miller (Chair of the Jury)
Natalie Mrachacz-Kersting (2017 Winner)
Vivek Prabhakaran
Yijun Wang
Milena Korostenskaja
Sharlene Flesher

This year's jury had an unusually high number of members who were nominated for BCI Research Awards in prior years. Like earlier juries, the 2018 jury included the preceding year's winner, Natalie Mrachacz-Kersting. The Chair of the jury, Kai Miller, was nominated in 2011 and 2014, and Milena Korostenskaja was nominated in 2017. Sharlene Flesher's team won second place in the 2016 BCI Research Awards for their work with an intracortical microstimulator as a feedback source for BCI users. Having a jury with alumni from previous awards should help the jury members appreciate the projects that they have to review.

The scoring criteria that the jury used to select the nominees and winners were the same as all previous BCI Research Awards:

- Does the project include a novel application of the BCI?
- Is there any new methodological approach used compared to earlier projects?
- Is there any new benefit for potential users of a BCI?
- Is there any improvement in terms of speed of the system (e.g. bit/min)?
- Is there any improvement in terms of accuracy of the system?
- Does the project include any results obtained from real patients or other potential users?
- Is the used approach working online/in real-time?
- Is there any improvement in terms of usability?
- Does the project include any novel hardware or software developments?

The jury then tallies the resulting scores, and the nominees are posted online. The nominees are invited to a Gala Awards Ceremony for that year's awards. For several years, this ceremony has been part of the biggest BCI conference for that year, which has had one of two organizers. The BCI Society organizes a BCI Meeting every even-numbered year, while the Technical University of Graz hosts a BCI Conference every odd-numbered year.

This year's ceremony was part of the 2018 BCI Meeting in Asilomar, CA. We held the ceremony outside, expecting a warm and dry evening, and the California weather did not disappoint anyone. Everyone enjoyed a pleasant beach breeze throughout the ceremony. We reviewed the BCI Awards process, asked the at least one nominee from each team to come receive a certificate and other prizes.Then, we announced

Fig. 1 The Chair of the Jury, Kai Miller, announces the projects nominated for the BCI Award at the BCI Meeting 2018. Christoph Guger is in the foreground on the right side of the picture, and Brendan Allison is behind him

the three winners and presented their prizes (Figs. 1 and 2). After the ceremony, many of us walked west to celebrate on the beach.

3 The BCI Research Award Book Series

The BCI Research Award Books have always been part of the BCI Research Awards. Since the first BCI Research Award in 2010, we published a book each year that presents the projects that were nominated this year. We encourage the authors of those chapters to add other material such as discussion and future directions.

Over the course of several years, we have decided to allow more time for authors to develop their contributions. This allows them time to publish underlying material and present new and more detailed information in their chapters, including additional results or discussion. For example, the next chapter from Mikhail Lebedev

Fig. 2 Christoph Guger (organizer), Kai Miller (Chair of Jury), Sharlene Flesher (jury), Josef Faller (nominee), Vivek Prabhakaran (jury), Brendan Allison (emcee)

and colleagues includes discussion about future directions to help patients with handwriting. In another example, the penultimate chapter from Chen Yang and colleagues presents details about how their BCI system works and how it was implemented with healthy persons and a patient.

After we receive the chapters from the authors, we review them, and sometimes request changes from the authors. After the chapters are ready, we send them to the publisher. The publisher sends back final proofs for review, and then the book goes into production.

Each year's book also begins with an introduction chapter like this one and ends with a chapter summarizing the highlights. Some books have included other material such as commentaries or, as is the case this year, interviews with authors. We've often noticed that the awards both track and anticipate changes in BCI research, and many chapters present emerging directions. The chapters introduce ideas, methods, and results that are relevant to readers who are interested in developing and building BCIs, as well as readers curious about basic scientific questions about how the brain works. Before these chapters from the nominees, the next section presents the nominated projects.

4   Projects Nominated for the BCI Award 2018

The jury scored all of the submissions using the scoring criteria (see Sect. 2 of this chapter), then tallied the scores to determine the twelve nominated for the BCI Research Award 2018. The nominees, including their affiliations and project names, were:

Alexei E. Ossadtchi[1*], Elizaveta Okorokova[2], Joseph S. Erlichman[3], Valery I. Rupasov[4], Mikhail A. Lebedev[1,5], and Michael Linderman[4]

Generating Handwriting from Multichannel EMG

1   National Research University Higher School of Economics.
2   University of Chicago.
3   St. Lawrence University.
4   Norconnect Inc.
5   Duke University.

Martin Burns, Dingyi Pei, Ramana Vinjamuri

Real-time EEG Control of a Dexterous Hand Exoskeleton embedded with Synergies

Department of Biomedical Engineering, Stevens Institute of Technology, NJ, USA.

James O'Sullivan[2], Zhuo Chen[1], Jose Herrero[4], Guy M McKhann[3], Sameer A Sheth[3], Ashesh D Mehta[4], Nima Mesgarani[1,2,5]

Neural decoding of attentional selection in multi-speaker environments without access to clean sources

1   Department of Electrical Engineering, Columbia University, New York, USA.
2   Mortimer B Zuckerman Mind Brain Behavior Institute, Columbia University, New York, USA.
3   Department of Neurological Surgery, The Neurological Institute, 710 West 168 Street, New York, USA.
4   Department of Neurosurgery, Hofstra-Northwell School of Medicine and Feinstein Institute for Medical Research, Manhasset, NY, USA.

J. Faller[1], J. Cummings[1], S. Saproo[1], P. Sajda[1,2]

BCI-based regulation of arousal improves human performance in a demanding sensory-motor task

1   Department of Biomedical Engineering, Columbia University, New York, USA.
2   Data Science Institute, Columbia University, New York, USA.

Christian Herff[1], Lorenz Diener[1], Emily Mugler[3], Marc Slutzky[3], Dean Krusienski[2], Tanja Schultz[1]

Brain-To-Speech: Direct Synthesis of Speech from Intracranial Brain Activity Associated with Speech Production

1 Cognitive Systems Lab, University of Bremen, Germany.
2 ASPEN Lab, Old Dominion University, Norfolk, USA.
3 Departments of Neurology, Physiology, and Physical Medicine & Rehabilitation, Northwestern University, Chicago, USA.

Abidemi Bolu Ajiboye[1,2,6], Francis R. Willett[1,2,6], Daniel R. Young[1,2,6], William D. Memberg[1,2,6], Brian A. Murphy[1,2,6], Jonathan P. Miller[2,4,6], Benjamin L. Walter[2,3,6], Jennifer A. Sweet[2,4,6], Harry A. Hoyen[5,6], Michael W. Keith[5,6], Paul Hunter Peckham[1,2,6], John D. Simeral[7,8,9,10], John P. Donoghue[8,9,12], Leigh R. Hochberg[7,8,9,10,11], Robert F. Kirsch[1,2,4,6]

Restoring Functional Reach-to-Grasp in a Person with Chronic Tetraplegia using Implanted Functional Electrical Stimulation and Intracortical Brain-Computer Interfaces

1 Department of Biomedical Engineering, Case Western Reserve University, Cleveland, Ohio, USA.
2 Louis Stokes Cleveland Department of Veterans Affairs Medical Center, FES Center of Excellence, Rehab. R&D Service, Cleveland, Ohio, USA.
3 Department of Neurology, University Hospitals Case Medical Center, Cleveland, Ohio, USA.
4 Department of Neurological Surgery, University Hospitals Cleveland Medical Center, Cleveland, Ohio, USA.
5 Department of Orthopaedics, MetroHealth Medical Center, Cleveland, Ohio, USA.
6 School of Medicine, Case Western Reserve University, Cleveland, Ohio, USA.
7 School of Engineering, Brown University, Providence, Rhode Island, USA.
8 Center for Neurorestoration and Neurotechnology, Rehabilitation R&D Service, Department of Veterans Affairs Medical Center, Providence, Rhode Island, USA.
9 Brown Institute for Brain Science, Brown University, Providence, Rhode Island, USA.
10 Department of Neurology, Massachusetts General Hospital, Boston, Massachusetts, USA.
11 Department of Neurology, Harvard Medical School, Boston, Massachusetts, USA.
12 Department of Neuroscience, Brown University, Providence, Rhode Island, USA.

Michael Tangermann[1,3], David Hübner[1,3], Simone Denzer, Atieh Bamdadian[4], Sarah Schwarzkopf[2,3], Mariacristina Musso[2,3]

A BCI-Based Language Training for Patients with Chronic Aphasia

1 Brain State Decoding Lab, Dept. Computer Science, Albert-Ludwigs-Universität Freiburg, Germany.
2 Department of Neurology, University Medical Center Freiburg, Germany.
3 Cluster of Excellence BrainLinks-BrainTools, Albert-Ludwigs-Universität Freiburg, Germany.
4 Inovigate, Aeschenvorstadt 55, 4051 Basel, Switzerland.

Robert Gabriel Lupu[1], Florina Ungureanu[1], Oana Ferche[2], Alin Moldoveanu[2]

Neuromotor Recovery based on BCI, FES, Virtual Reality and Augmented Feedback for upper limbs

1 Computer Engineering Department, "Gheorghe Asachi" Technical University of Iasi, Romania.
2 Computer Engineering Department, "Politehnica" University of Bucharest, Romania.

Chen Yang, Xiang Li, Shangkai Gao, Xiaorong Gao

A Dynamic Window SSVEP-Based Brain-Computer Interface System using a Spatio-Temporal Equalizer

Department of Biomedical Engineering, Tsinghua University Beijing, P.R. China.

S. Perdikis, L. Tonin, S. Saeedi, C. Schneider, J. del R. Millán

Successful mutual learning with two tetraplegic users: The Cybathlon BCI race experience

Defitech Chair in Brain-Machine Interface (CNBI), École Polytechnique Fédérale de Lausanne (EPFL), Geneva, Switzerland.

Andrew G. Richardson[1], Yohannes Ghenbot[1], Xilin Liu[2], Han Hao[2], Sam DeLuccia[1], Gregory Boyek[1], Solymar Torres-Maldonado[1], Firooz Aflatouni[2], Jan Van der Spiegel[2], Timothy H. Lucas[1]

A Wireless Sensory Interface to Inform Goal-Directed Actions

1 Department of Neurosurgery, University of Pennsylvania, Philadelphia, PA, USA.
2 Department of Electrical and Systems Engineering, University of Pennsylvania, Philadelphia, PA, USA.

S. Perdikis, S. Saeedi, J. del R. Millán

**Longitudinal training and use of non-invasive motor imagery BCI
by an incomplete locked-in user**

Defitech Chair in Brain-Machine Interface (CNBI), École Polytechnique Fédérale de Lausanne (EPFL), Geneva, Switzerland.

5 Summary

As BCI research continues to expand, new directions are becoming increasingly prevalent. Novel BCI approaches for handwriting support, therapy, prosthetic control, speech production, and hearing aids were among the nominees this year, any or all of which could lead to new devices for patients. These directions have also gained attention in the broader BCI research community, and the following chapters present some of the best recent work that's pushing our field forward. The discussion chapter announces and interviews this year's winners, followed by our concluding comments.

Generating Handwriting from Multichannel Electromyographic Activity

Mikhail A. Lebedev, Alexei E. Ossadtchi, Elizaveta Okorokova, Joseph S. Erlichman, Valery I. Rupasov, and Michael Linderman

Abstract Handwriting is an advanced motor skill and one of the key developments in human culture. Here we show that handwriting can be decoded—offline and online—from electromyographic (EMG) signals recorded from multiple hand and forearm muscles. We convert EMGs into continuous handwriting traces and into discretely decoded font characters. For this purpose, we use Wiener and Kalman filters, and machine learning algorithms. Our approach is applicable to clinical neural prostheses for restoration of dexterous hand movements, and to medical diagnostics of neural disorders that affect handwriting. We also propose that handwriting could be decoded from cortical activity, such as the activity recorded with electrocorticography (ECoG).

Keywords EMG · ECoG · Handwriting · BCI · Neural prosthetics

1 Introduction

The development of bioelectric interfaces holds significant promise for both clinical and consumer applications [11]. Although handwriting is one of the essential motor skills, it has received relatively little attention from the developers of both brain-computer and myoelectric interfaces. This is because the major focus of the

M. A. Lebedev · A. E. Ossadtchi (✉)
National Research University Higher School of Economics, Moscow, Russia
e-mail: ossadtchi@gmail.com

M. A. Lebedev
e-mail: mikhail.a.lebedev@gmail.com

E. Okorokova
University of Chicago, Chicago, USA

J. S. Erlichman
St. Lawrence University, New York, USA

V. I. Rupasov · M. Linderman
Norconnect Inc., New York, USA

© The Author(s), under exclusive license to Springer Nature Switzerland AG 2020
C. Guger et al. (eds.), *Brain–Computer Interface Research*,
SpringerBriefs in Electrical and Computer Engineering,
https://doi.org/10.1007/978-3-030-49583-1_2

research on brain-computer interfaces (BCIs) [2, 4, 12] and myoelectrical interfaces [14, 15, 26] has been on arm reaching and grasping movements. Here we report a research program aimed at the development of bioelectric interfaces for decoding of handwriting patterns from EMG signals and brain activity.

We have demonstrated that surface EMGs recorded from the forearm and hand muscles can be translated into pen traces exhibited during handwriting or discrete font characters. EMG-based interfaces can be utilized in a similar fashion as BCIs that convert brain activity into hand coordinates [2, 12]. Additionally, EMG-based interfaces could be convenient for testing decoding algorithms prior to their utilization in BCIs, which are harder to implement. Furthermore, we envision several EMG-based approaches to functional restoration and/or augmentation, such as myoelectrical prostheses for amputees (e.g., a myoelectrically driven prosthetic hand) and for healthy people (e.g., an EMG glove instead of a keyboard).

EMG-based interfaces that generate handwriting could be used in clinics as well. Indeed, handwriting often deteriorates in neurological conditions, including dementia [17], Parkinson's disease [28], writing tremor [6], and attention deficit hyperactivity disorder [20]. Accordingly, EMG-based interfaces could be utilized for diagnostics of these disorders and monitoring of their treatment. Overall, we suggest that interfaces that decode handwriting could provide both fundamental insights on neurophysiological mechanisms of fine hand movements and useful clinical approaches for a range of neural diseases.

2   Decoding EMG with the Wiener Filter

We pioneered an EMG-based interface for decoding handwritten patterns [16]. In this study, we developed two main approaches for such decoding. In the first approach, we continuously extracted pen traces from EMGs using the Wiener filter as decoding algorithm. In the second approach, we decoded font characters using a discrete classifier. Both approaches are applicable to many types of BCIs, for example ECoG-based interfaces.

Figure 1 shows the experimental setup of this study. Eight bipolar surface EMG electrodes were placed over the muscles of the forearm (flexor carpi radialis, extensor digitorum, extensor carpi ulnaris, extensor carpi radialis) and intrinsic hand muscles (opponens pollicis, abductor pollicis brevis, and medial and lateral heads of first dorsal interosseous). The electrical activity of these muscles exhibited clear EMG modulations when human subjects performed handwriting. We quantified these modulations as changes in rectified EMG (full-wave rectification followed by low-pass filtering with a 5 Hz cutoff). The decoding step consisted of the Wiener filters expressing X and Y coordinates of the pen as weighted sums of the rectified EMG measurements (Fig. 2a):

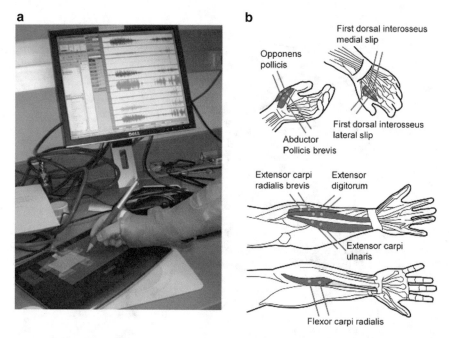

Fig. 1 Decoding of handwriting from EMG. **a** Experimental setup. **b** Placement of surface EMG electrodes over hand and forearm muscles. Reproduced from Linderman et al. [16]

$$x(t) = b + \sum_{\Delta t=-T}^{T} \mathbf{w}^T(\Delta t)\mathbf{n}(t + \Delta t) + \varepsilon(t) \tag{1}$$

where $x(t)$ is the pen coordinate, t is time, $\mathbf{n}(t + \Delta t)$ is a vector containing multichannel EMG data at time t and time-shift Δt, $2T+1$ is the analysis time-window, $\mathbf{w}(\Delta t)$ is a vector of weights, b is the y-intercept, and $\varepsilon(t)$ is the residual error. The weights were calculated in MATLAB using the function *regress*.

The results of decoding with the Wiener filter are shown in Fig. 2b. Blue lines correspond to the actual pen traces and red lines correspond to pen traces extracted from multichannel EMG data. While the Wiener filter is a relatively simple decoding algorithm, it performs well [16], with decoding accuracy comparable to the accuracy of invasive BCIs [2, 12]. The decoding accuracy improved when the number of EMG channels increased, which is consistent with the results of multichannel BCI studies [2, 11, 12].

In a follow-up study, we determined that EMG modulations exhibited during handwriting were well approximated by a lognormal distribution [22]. This suggests that taking lognormal distribution into account could aid the development of better algorithms for EMG decoding.

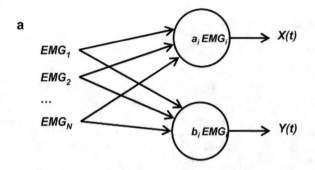

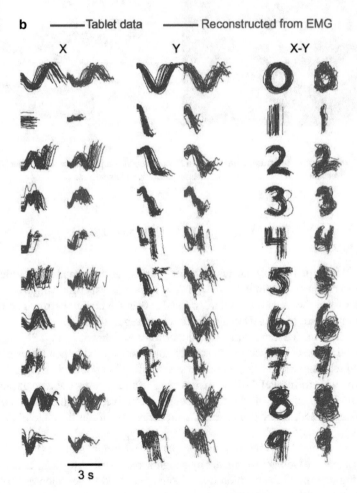

Fig. 2 Decoding with the Wiener filter. **a** Schematics of the Wiener filter. **b** Decoded pen. Actual traces are shown in blue; reconstructed traces are shown in red. Reproduced from Linderman et al. [16]

In addition to converting EMG into pen traces offline, we conducted the same Wiener filter-based conversion online (unpublished observations). In these experiments, pen traces were generated while subjects wrote characters (digits or letters) on a computer tablet.

3 Classifying EMGs Into Font Characters

A range of practical applications, such as EMG-based keyboards, could utilize discrete classification of font characters from EMG instead of continuously reconstructing pen traces. To assess the possibility of such decoding, we developed a discrete-classification approach.

For discrete classification, we utilized Fisher linear discriminant analysis (LDA; MATLAB function *classify*) that converted multichannel EMG data into font characters ("0", "1", ... "9"; Fig. 3). To trigger classification, the onset of handwriting was detected for each character. The onset was computed using the compound EMG, i.e. the sum of rectified (and converted to z-scores) EMGs for all channels. The onset of handwriting was defined as the compound EMG crossing the threshold equal to 0.5 standard deviations from the background value. Next, the EMG time series were subdivided into 3.5-s epochs corresponding to the writing of each character. EMG templates were calculated for each muscle and each character. The templates were the across-trial average traces of the rectified EMGs. These templates were slid along the EMG time series, and correlation between the EMGs and the templates was continuously evaluated using correlation coefficient, R, as the metric. Peak R values that occurred when the templates were well aligned with the writing segment were used for LDA-based classification.

Ten-character classification was performed with this method. For six subjects that participated in the study, classification accuracy was $90.4 \pm 7.0\%$ (mean $\pm$ standard deviation across subjects). The accuracy was lower if only the intrinsic hand muscles ($79.2 \pm 10.6\%$) or only the forearm muscles ($83.5 \pm 10.0\%$) were used for classification, and it was the lowest if only one muscle was used ($51.6 \pm 12.5\%$).

A potentially useful approach to transform EMG classification into font characters would include a combination of continuous reconstruction of pen traces and a discrete classifier. In this approach, EMGs would be converted into a pen trace first and the reconstructed pen trace would be then processed by one of the available algorithms for handwriting recognition [8, 19]. This idea should be tested experimentally in the future.

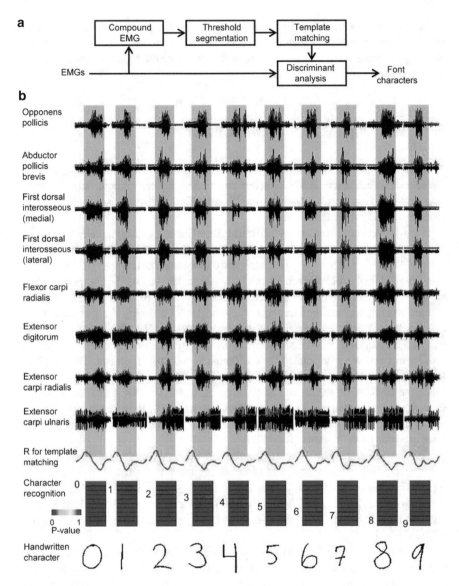

Fig. 3 Reconstruction of font characters using linear discriminant analysis. Reproduced from Linderman et al. [16]

4 Decoding EMGs with the Kalman Filter

To improve the decoding obtained with the Wiener filter, we developed a decoder based on the Kalman filter (KF) [18], the algorithm that fuses several noisy signals to generate an estimate of a dynamical system's state, which minimizes the squared error

[10]. In the case of handwriting, the Signal 1 is derived from the dynamical model of the hand holding the pen. This model is formalized as a multivariate autoregressive process that can be estimated from the pen data. Signal 2 is the vector of multichannel EMG. The relationship between these two signals is modeled using a multivariate linear regression derived from the training dataset.

The changes in Signal 1, which we call state vector, are described by the equation:

$$\mathbf{s}_t = \mathbf{A}\mathbf{s}_{t-1} + \mathbf{v}_t \tag{2}$$

where $\mathbf{s}_t$ is the state vector that contains pen coordinates and their first and second derivatives, $\mathbf{A}$ is state transition matrix, and $\mathbf{v}_t$ is the vector containing noise.

In the KF implementations, the measurement equation is often written as $\mathbf{z} = F(\mathbf{s})$, where $\mathbf{z}$ is Signal 2, or the vector of measurements. However, in the case of handwriting, it is more natural to assume that EMGs produce pen movements. Therefore, we used the inverse form of this equation as the measurement model:

$$\mathbf{s}_t = \mathbf{H}\mathbf{s}_{t-1} + \mathbf{w}_t \tag{3}$$

where z_t is the vector containing multichannel EMG measurements, $\mathbf{H}$ is measurement transformation, and $\mathbf{w}_t$ is the vector of measurement noise.

To reconstruct the state vector using both signals, we performed a statistical fusion of the estimates based on the state and the measurement models:

$$\boldsymbol{\mu}_{fused} = (1 - \mathbf{K}_t)\boldsymbol{\mu}_{1t} + \mathbf{K}_t\boldsymbol{\mu}_{2t} \tag{4}$$

where $\boldsymbol{\mu}_{fused}$ is a weighted sum of the two mean vectors of the signals, and $\mathbf{K}_t$ is the Kalman gain that serves as the dynamic scaling factor reflecting the relative trust in each of the two signals.

We used two designs for decoding of handwriting with the KF. In the generic design, a single set of parameters was calculated using the training data for all written characters (from "0" to "9"). In the specific design, a separate set of parameters was calculated for each character. We assessed the performance of these approaches when predictive, i.e. only EMG data from the past, was used. For the generic design, we achieved the average across all subjects trajectory reconstruction accuracy of $63 \pm 17\%$ and $73 \pm 14\%$, for X and Y coordinates, respectively, which exceeded the performance of the Wiener filter (47 ± 2 and $63 \pm 15\%$), (Fig. 4). For the specific design, we achieved an even better accuracy ($78 \pm 13\%$ and $88 \pm 7\%$). Interestingly, the decoder trained on the odd digits performs well on the even digits, which may speak about the generalization achieved by the described approach.

Thus, the method worked well for both specific and general models. While the application of the generic design for practical decoding of handwriting is straightforward, the specific design needs additional tools to become practical. This is primarily related to the fact that when dealing with amputees, no training data in the form of recorded pen traces are available and new tools for training both the user and the decoder need to be developed. Within this approach, after some preliminary training,

Fig. 4 Single trial pen trace reconstruction of digits zero to nine with the Kalman Filter (left) and the Wiener filter (right), several cross-validation trials of one of the participants

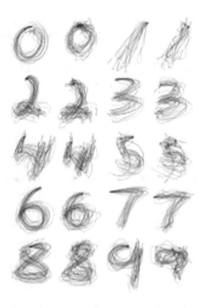

we can use a discrete classifier applied first to classify characters into groups, followed by a specific KF that generates pen traces. One approach to build such a classifier can be based on the Hidden Markov Model machinery.

5 Hidden Markov Model for Discrete Classification

In addition to decoding handwriting with the KF, we developed a hidden Markov model (HMM) that described handwriting as a series of state transitions. Each figure is represented in the HMM as a specific sequence of states. This is quite clearly seen in Fig. 5, where each color corresponds to a specific state. According to our assessment (unpublished observations), HMM was a very efficient method for decoding handwritten patterns. An accuracy up to 97% was reached.

Electrocorticographic recordings during handwriting tasks We have started a series of experiments in which electrocorticographic (ECoG) activity is recorded in human patients that perform motor tasks with a hand-held pen (Fig. 6). As a starting point, we chose to use a center-out task originally introduced by Georgopoulos and his colleagues to investigate directional tuning properties of monkey motor cortical neurons [7]. Similar experimental paradigms were utilized to characterize directional properties of ECoG signals recorded in humans [1, 9, 13, 21, 23, 27].

In our study, we implement a version of the center-out paradigm in which the pointer was controlled by the subject moving a pen on the surface of the digitizing tablet. Thus, the task was very similar to writing with a pen. This paradigm can be extended in the future to a range of drawing and handwriting tasks.

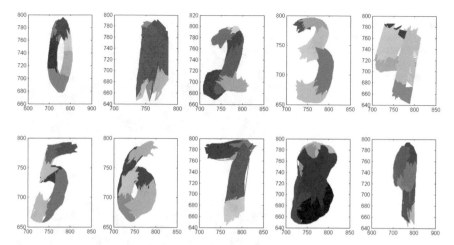

Fig. 5 Traces of handwritten characters marked with colors representing HMM states. Each character, thus, corresponds to a specific path in the HMM graph and can be decoded with 97% accuracy

To explore the time evolution of the directionally tuned ECoG features, R^2 values were calculated for different ECoG frequency bands and for different time points with respect to the center-task events (like go-cue or movement onset). We used single feature-based regression of movement direction. R^2 values were represented as heat maps (Fig. 6). For the patients with implanted grids, spatial coherence (i.e. continuity of regions in time and frequency that yield higher correlation) can be seen. The highest correlation values were observed during the pre-movement and movement periods. R^2 ranged 0.21–0.42 during the target presentation period and 0.25–0.41 during the movement period. Directional tuning was especially prominent in the gamma band during target presentation in all patients and in some patients after movement onset.

Based on our preliminary results, ECoG appears to be a useful signal for decoding pen traces. We foresee that ECoG recordings could potentially provide insights on cortical mechanisms of handwriting and could be utilized in BCIs that generate handwriting from cortical activity.

6 Discussion

Handwriting is a unique human skill made possible by evolution and cultural development. Even though modern technological developments require learning of different motor skills, such as typing on a tablet, handwriting remains one of the primary means of communicating and recording. Handwriting is hard to replicate in prosthetic devices, particularly when a control by bioelectrical signals is required.

Fig. 6 R^2 maps for the three frequency bands of ECoG for the center-out task. The bottom plot shows the average pen trajectory

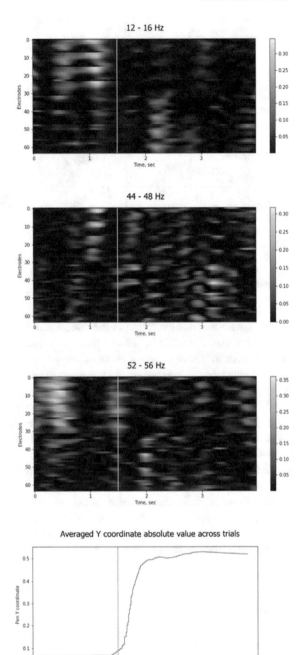

Generating handwriting from the EMGs of hand muscles (and possibly other muscles of the body, which could be a solution for amputees or partially paralyzed patients) is a difficult problem because of the variability of EMGs during natural handwriting and their dependence on the posture. In our experiments, we observed that changes in the way the pen was held caused significant alterations in the EMG patterns exhibited during handwriting, even though pen traces changed very little. This is an example of motor redundancy, where very different muscle activation patterns could produce the same trajectory of the limb. Because of this variability of EMG patterns, extraction of handwriting from EMG is more difficult than computer recognition of handwritten characters. Experimentally, we handled this problem by asking the subjects to hold the pen the same way and to maintain the same speed of handwriting. This requirement could be harder to fulfill in real-life situations. Yet, there is hope that modern machine learning algorithms, such as deep learning, could handle this problem by developing efficient solutions based on a large number of examples.

In our work on decoding of handwriting from EMG, we have probed several continuous and discrete decoders and observed good decoding accuracy. We concluded that computer algorithms can convert multichannel EMG into pen traces and/or font characters, particularly when the character set is relatively small (ten characters in our experiments). We found that the decoding performance improved when more advanced algorithms were used, which suggests that there is still room for improvement in these methods and practical EMG-based generators of handwriting are feasible for both patients and healthy subjects.

With the current hype around BCIs, EMG-based interfaces have received less attention. Yet, such interfaces are of considerable interest for several reasons. One reason is the simplicity of obtaining high-quality EMG recordings. The signal-to-noise ratio (SNR) of EMG recordings by far exceeds the SNR of EEG and even ECoG recordings. Although one might argue that EMG is not a true brain signal, this signal does replicate the discharges of spinal motoneurons, so it is clearly a neural signal. Overall, we think that EMG-based interfaces could be even considered as a subtype of BCIs. Importantly, EMG-based interfaces could be much more practical than BCIs in many cases. Indeed, BCI developers treat EMGs as an artifact that hinders recording of low-amplitude brain activity. Yet, EMGs could be of great practical value, whereas the value of using pure brain activity is questionable in many cases. For instance, one study reported that EMGs picked by the EEG electrodes yield a better control signal than the EEG purified from the EMG artifacts [5]. In many clinical cases, paralysis is partial. For example, paraplegic patients retain mobility in the arms. In such cases, EMG electrodes could be attached to mobile parts of the body to provide a high-quality control signal for a prosthetic device. Finally, since EMG signals bear similarity to neuronal activity of the brain, they can be utilized as an inexpensive testbed for various BCI designs.

We foresee that the development of EMG-based interfaces, including the ones that produce handwriting, will contribute to both practical applications and basic neurophysiology of motor control. This is because a good understanding of how EMG patterns are converted into hand movements will shed light on how vast brain networks generate delicate movements and what goes wrong with this control in neurological conditions. Clinical EMG-based approaches are of particular interest as they are applicable to rehabilitation of patients with a loss of a limb [30], handwriting-based diagnostics [24, 25, 29], and the development of writing skills in children [3].

As to the ECoG-based approach, much remains to be seen. At this point, we would speculate that the analysis of ECoG activity sampled from multiple cortical areas could clarify the conversion of language into written text, from the representation of thoughts to generation of motor commands sent to the spinal cord for execution. A better understanding of this transformation will result in a new generation of BCIs enabling handwriting and neurofeedback-based approaches to neurological disorders.

References

1. T. Ball, A. Schulze-Bonhage, A. Aertsen, C. Mehring, Differential representation of arm movement direction in relation to cortical anatomy and function. J. Neural Eng. **6**(1), 016006 (2009)
2. J.M. Carmena, M.A. Lebedev, R.E. Crist, J.E. O'Doherty, D.M. Santucci, D.F. Dimitrov, P.G. Patil, C.S. Henriquez, M.A. Nicolelis, Learning to control a brain–machine interface for reaching and grasping by primates. PLoS Biol. **1**(2), e42 (2003)
3. J.L. Carter, H.L. Russell, Use of EMG biofeedback procedures with learning disabled children in a clinical and an educational setting. J. Learn. Disabil. **18**(4), 213–216 (1985)
4. J.L. Collinger, B. Wodlinger, J.E. Downey, W. Wang, E.C. Tyler-Kabara, D.J. Weber, A.J. McMorland, M. Velliste, M.L. Boninger, A.B. Schwartz, High-performance neuroprosthetic control by an individual with tetraplegia. The Lancet **381**(9866), 557–564 (2013)
5. Á. Costa, E. Hortal, E. Iáñez, J.M. Azorín, A supplementary system for a brain-machine interface based on jaw artifacts for the bidimensional control of a robotic arm. PLoS ONE **9**(11), e112352 (2014)
6. R.J. Elble, C. Moody, C. Higgins, Primary writing tremor. A form of focal dystonia? Movem. Disord. Off. J. Movem. Disord. Soc. **5**(2), 118–126 (1990)
7. A.P. Georgopoulos, J.F. Kalaska, R. Caminiti, J.T. Massey, On the relations between the direction of two-dimensional arm movements and cell discharge in primate motor cortex. J. Neurosci. **2**(11), 1527–1537 (1982)
8. A. Graves, J. Schmidhuber, Offline handwriting recognition with multidimensional recurrent neural networks. Adv. Neural Inf. Process. Syst. (2009)
9. A. Gunduz, P. Brunner, M. Sharma, E.C. Leuthardt, A.L. Ritaccio, B. Pesaran, G. Schalk, Differential roles of high gamma and local motor potentials for movement preparation and execution. Brain-Comput. Interf. **3**(2), 88–102 (2016)
10. R.E. Kalman, A new approach to linear filtering and prediction problems. J. Basic Eng. **82**(1), 35–45 (1960)
11. M. Lebedev, Brain-machine interfaces: an overview. Transl. Neurosci. **5**(1), 99–110 (2014)
12. M.A. Lebedev, J.M. Carmena, J.E. O'Doherty, M. Zacksenhouse, C.S. Henriquez, J.C. Principe, M.A. Nicolelis, Cortical ensemble adaptation to represent velocity of an artificial actuator controlled by a brain-machine interface. J. Neurosci. **25**(19), 4681–4693 (2005)
13. E.C. Leuthardt, G. Schalk, J.R. Wolpaw, J.G. Ojemann, D.W. Moran, A brain–computer interface using electrocorticographic signals in humans. J. Neural Eng. **1**(2), 63 (2004)

14. M.V. Liarokapis, P.K. Artemiadis, P.T. Katsiaris, K.J. Kyriakopoulos, E.S. Manolakos Learning human reach-to-grasp strategies: towards EMG-based control of robotic arm-hand systems, in *2012 IEEE International Conference on Robotics and Automation, IEEE* (2012)

15. M.V. Liarokapis, P.K. Artemiadis, K.J. Kyriakopoulos, E.S. Manolakos, A learning scheme for reach to grasp movements: on EMG-based interfaces using task specific motion decoding models. IEEE J. Biomed. Health Inf. **17**(5), 915–921 (2013)

16. M. Linderman, M.A. Lebedev, J.S. Erlichman, Recognition of handwriting from electromyography. PLoS ONE **4**(8), e6791 (2009)

17. C. Luzzatti, M. Laiacona, D. Agazzi, Multiple patterns of writing disorders in dementia of the Alzheimer type and their evolution. Neuropsychologia **41**(7), 759–772 (2003)

18. E. Okorokova, M. Lebedev, M. Linderman, A. Ossadtchi, A dynamical model improves reconstruction of handwriting from multichannel electromyographic recordings. Front. Neurosci. **9**, 389 (2015)

19. R. Plamondon, S.N. Srihari, Online and off-line handwriting recognition: a comprehensive survey. IEEE Trans. Pattern Anal. Mach. Intell. **22**(1), 63–84 (2000)

20. M.B. Racine, A. Majnemer, M. Shevell, L. Snider, Handwriting performance in children with attention deficit hyperactivity disorder (ADHD). J. Child Neurol. **23**(4), 399–406 (2008)

21. C.G. Reddy, G.G. Reddy, H. Kawasaki, H. Oya, L.E. Miller, M.A. Howard, Decoding movement-related cortical potentials from electrocorticography. Neurosurg. Focus **27**(1), E11 (2009)

22. V.I. Rupasov, M.A. Lebedev, J.S. Erlichman, M. Linderman, Neuronal variability during handwriting: lognormal distribution. PLoS ONE **7**(4), e34759 (2012)

23. J.C. Sanchez, A. Gunduz, P.R. Carney, J.C. Principe, Extraction and localization of mesoscopic motor control signals for human ECoG neuroprosthetics. J. Neurosci. Methods **167**(1), 63–81 (2008)

24. M.C. Silveri, F. Corda, M.N. Di, Central and peripheral aspects of writing disorders in Alzheimer's disease. J. Clin. Exp. Neuropsychol. **29**(2), 179–186 (2007)

25. V. Stanford, Biosignals offer potential for direct interfaces and health monitoring. IEEE Pervasive Comput. **3**(1), 99–103 (2004)

26. J. Tian, J. He, Can EMG machine interface be used to model brain machine interface? in *Proceedings of the 25th Annual International Conference of the IEEE Engineering in Medicine and Biology Society* (IEEE Cat. No. 03CH37439), IEEE (2003)

27. C. Toro, C. Cox, G. Friehs, C. Ojakangas, R. Maxwell, J.R. Gates, R.J. Gumnit, T.J. Ebner, 8–12 Hz rhythmic oscillations in human motor cortex during two-dimensional arm movements: evidence for representation of kinematic parameters. Electroencephalography Clin. Neurophys. Evoked Potent. Sect. **93**(5), 390–403 (1994)

28. O. Tucha, L. Mecklinger, J. Thome, A. Reiter, G. Alders, H. Sartor, M. Naumann, K.W. Lange, Kinematic analysis of dopaminergic effects on skilled handwriting movements in Parkinson's disease. J. Neural Trans. **113**(5), 609–623 (2006)

29. A. Van Gemmert, H.-L. Teulings, J.L. Contreras-Vidal, G. Stelmach, Parkinsons disease and the control of size and speed in handwriting. Neuropsychologia **37**(6), 685–694 (1999)

30. Z.G. Xiao, C. Menon, Towards the development of a wearable feedback system for monitoring the activities of the upper-extremities. J. Neuroeng. Rehabil. **11**(1), 2 (2014)

Neural Decoding of Upper Limb Movements Using Electroencephalography

Dingyi Pei, Martin Burns, Rajarathnam Chandramouli, and Ramana Vinjamuri

Abstract Rationale: The human central nervous system (CNS) effortlessly performs complex hand movements with the control and coordination of multiple degrees of freedom (DoF), but how those mechanisms are encoded in the CNS remains unclear. In order to investigate the neural representations of human upper limb movement, scalp electroencephalography (EEG) was recorded to decode cortical activity in reaching and grasping movements. **Methods**: Upper limb movements including arm reaching and hand grasping tasks were observed in this study. EEG signals of 15 healthy individuals were recorded (g.USBamp, g.tec, Austria) when performing reaching and grasping tasks. Spectral features of the relevant cortical activities were extracted from EEG signals to decode the relevant reaching direction and hand grasping information. Upper limb motion direction and hand kinematics were captured with sensors worn on the hands. Directional EEG features were classified using stacked autoencoders; hand kinematic synergies were reconstructed to model the relationship of hand movement and EEG activities. **Results**: An average classification accuracy of three-direction reaching tasks achieved 79 $\pm$ 5.5% (best up to 88 $\pm$ 6%). As for hand grasp decoding, results showed that EEG features were able to successfully decode synergy-based movements with an average decoding accuracy of 80.1 $\pm$ 6.1% (best up to 93.4 $\pm$ 2.3%). **Conclusion**: Upper limb movements, including directional arm reaching and hand grasping expressed as weighted linear combinations of synergies, were decoded successfully using EEG. The proposed decoding and control mechanisms might simplify the complexity of high dimensional motor control and might hold promise toward real-time neural control of synergy-based prostheses and exoskeletons in the near future.

Keywords Brain computer interface (BCI) · Electroencephalography (EEG) · Arm reaching movements · Hand kinematic synergies

D. Pei · M. Burns · R. Chandramouli · R. Vinjamuri (✉)
Sensorimotor Control Laboratory, Stevens Institute of Technology, Hoboken, NJ, USA
e-mail: ramana.vinjamuri@stevens.edu

© The Author(s), under exclusive license to Springer Nature Switzerland AG 2020

C. Guger et al. (eds.), *Brain–Computer Interface Research*,
SpringerBriefs in Electrical and Computer Engineering,
https://doi.org/10.1007/978-3-030-49583-1_3

25

Brain computer interfaces (BCIs) have become increasingly popular in recent years, bridging the gap between neural representations and external devices. Significant applications of BCI systems in motor control have been successfully accomplished [1]. Motor control aims to decode movement intention and execution encoded in multiple cortical areas, involving the integration and coordination of sensory and cognitive information. The neural signals can be decoded to trigger assistive devices to translate user intention into prosthetic control, or to drive an exoskeleton to assist and promote the user's natural movements in rehabilitation applications [2].

A simple hand movement includes arm reaching and hand grasping. In total, this involves simultaneously coordinating around 40 degrees of freedom (DoFs), requiring shoulder and elbow extension/flexion, wrist extension/flexion/rotation, and finger and thumb extension/flexion. Furthermore, each of the 28 DoFs of the hand are coordinated synchronously to achieve a simple reaching and grasping task. The central nervous system (CNS) may control every single digit individually, but it is difficult to comprehend the high complexity of control at the neural level. Currently, the main challenge for BCIs is to decode the kinematics of the upper limb from brain signals. Simplified methods have been proposed where high numbers of individual DoFs may be recruited with lower dimensional vectors in the CNS [3–5]. Motor decoding approaches that use invasive intracortical techniques achieve high performance but are mostly applied to non-human primates. Non-invasive methods such as Electroencephalography (EEG), however, are widely used in humans and are good candidates for studying motor control despite their limitations due to low signal to noise ratio. Our study aims to find optimal computational approaches to decode upper limb movements from scalp EEG.

1 Reaching and Grasping

1.1 Directional Arm Reaching

Reaching is a complex task that involves integration of information from upper limb kinematics, coordinating with target direction and reference frame transformation [6]. How the CNS integrates and coordinates this information is an intriguing topic in research. Movement execution and eye movement coordination are performed by the cortical networks in the frontal and parietal cortices [7] occurring in the early movement stage, approximately 1.5 s before motor action. The posterior parietal cortex organizes the transformations between different reference frames in the movement planning phase [8]. Additionally, hand motions and reaching orientations are processed and integrated in the dorsolateral prefrontal cortex and dorsal premotor cortex [9].

A majority of the neuromotor research in upper limb reaching has investigated the prediction of movement direction [10], detection of early neural correlates of movement [10, 11], and decoding the end point in reaching from neural recordings

[12]. Most reaching studies, such as center-out tasks, were based on synchronous neural responses [13]. Studies demonstrated that asynchronous reaching movements can better decode natural motor directional modulation in humans [14], where slow cortical potentials (SCP, <1 Hz) were widely used. This readiness potential is an unconscious preparatory brain activity that occurs about one second before the movement, preceding the conscious awareness to act [15]. Studies showed that the brain is activated 1.5–2 s before the voluntary movement execution. In particular, self-paced wrist extension onset can be detected around 0.62 s before actual movement [11].

There are some challenges in decoding reaching movement using EEG based BCI systems. The user must be fully concentrating during the experiment, since unwanted activity from irrelevant neural activations will lead to unexpected signal patterns [16]. Another challenge is the difficulty in separating control-state and idle-state signals especially in asynchronous tasks. Idle signals would cause false positives in classification when the user is not focused on the task [17].

1.2 Hand Grasp Kinematics

The high number of DoFs of the hand makes motor decoding difficult for hand grasping. So far, the mechanism of how the high DoFs of hand movement are encoded in the CNS needs further investigation. Previous studies proposed simplified approaches of recruiting hand movements with lower dimensional vectors. It is hypothesized the CNS may work in a higher level control rather than coordinate individual DoF in the lower level [3–5, 18]. Nikolai Bernstein [3] proposed the idea of synergy-based movement, indicating that the CNS adapts simplified strategies using global variables to reduce the complexity in motor control. Synergy control provides a simplified, higher level concept compared with independent DoFs control, with the internal functional integrity of individual digits preserved [19]. Evidence suggests that synergy-based movement is not just a theoretical complexity reduction approach but could be used as an optimal strategy by the CNS in simplifying and achieving complicated movements [20]. Nevertheless, the anatomical location of hand synergies has not yet been identified. The main challenge is exploring the correlations between synergies and the corresponding encoded neural representations [21].

Studies based on non-human primates indicated that hand movement information, such as hand velocity, is embedded in the frontal, parietal cortex, and cerebellum [22, 23]. In the hand joint kinematic investigation, it was found that the first principal component accounted for approximately 70–95% of variance in hand grasping movements [24, 25]. Each principal component is regarded as a kinematic synergy. These synergies support the majority of grasp types in daily life. It is evident that synergy formation and representation occur in multiple brain cortices. Thus, this study characterizes various cortical brain regions in order to further the understanding of synergy control at the global cortical activity level using EEG.

2 Methods

2.1 Subjects and Data Acquisition

Fifteen naive untrained subjects (eight males, seven females, aged 23 ± 3.1, five for experiment I and ten for experiment II) were recruited in this experiment under Stevens Institute of Technology Institutional Review Board approval.

Upper limb movement is integrated by visual and somatomotor information formed in the posterior parietal cortex and processed by the frontal cortex. This information is then sent to premotor cortex to coordinate reaching and grasping behavior [8, 26]. In our experiment, EEG was continuously recorded from those areas by an EEG cap (g.GAMMA cap, g.tec, Schiedlberg, Austria) and two amplifiers (g.USBamps, g.tec, Schiedlberg, Austria) using BCI2000 [27] software at a sampling rate of 256 Hz. 32 electrodes (F3, F1, Fz, F2, F4, FC5, FC3, FC1, FCz, FC2, FC4, FC6, C5, C3, C1, Cz, C2, C4, C6, CP5, CP3, CP1, CPz, CP2, CP4, CP6, P3, P1, Pz, P2, P4 and POz) were used for recording the signals, with impedance kept below 5 kΩ.

Hand trajectory and joint kinematic data was recoded using a motion tracker (Liberty, Polhemus, Vermont, USA) and CyberGlove worn on the subjects' right hand respectively. In this study, 10 of 18 sensors was used to measure the metacarpophalangeal (MCP) joints and proximal interphalangeal (PIP) joints of the fingers (IP for thumb). Each subject performed initial postures to calibrate the glove. CyberGlove data was captured at 125 Hz using a custom-built LabVIEW (National Instruments Corporation, Austin, TX, USA) program.

2.2 Experimental Protocol and Data Analysis

Experiment I—Three-directional Reaching: This experiment aims to determine the forearm reaching direction from brain activity [28]. Subjects were asked to sit in front of the experiment table to conduct a self-paced three-directional forearm reaching task, as shown in Fig. 1a. Three targets are located in three directions (left, central and right). In each trial, subjects were asked to focus on the target for about 2 s before the actual movement, reach the target at their own pace, hold the target for about 2 s, and then return to the start position.

The recorded raw EEG data was first re-referenced to the global potential, and then filtered into eight bandwidths (1–4, 4–8, 6–16, 13–30, 30–45, 1–45, 8–30 and 70–100 Hz). Autoregressive power spectral densities (PSDs) were estimated with selected EEG segment from 1.5 s before to 3.5 s after the forearm movement. Stacked autoencoders are utilized on directional classification. The traditional EEG classification methods, principal component analysis and linear discriminant analysis (PCA-LDA) were also applied here for comparison.

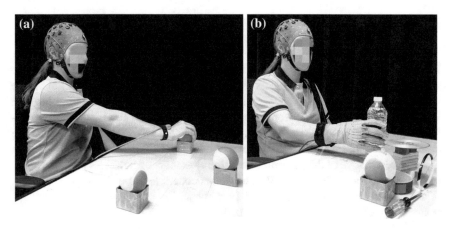

Fig. 1 Experiment setup. Three-direction reaching task (**a**) and six-posture hand grasping task (**b**)

Experiment II—Six-posture Hand Grasping: The goal of this experiment is to determine the correlations between brain motor activity and hand kinematics [29, 30]. Unlike in Experiment I, the hand grasping task is conducted in one reaching direction to perform six grasp types (tripod, cylindrical, lateral, spherical, hook and precision, Fig. 1b) found in activities of daily life.

EEG features are analyzed in a similar way as that of Experiment I, and synergies were extracted from hand kinematics using singular value decomposition (SVD). We assume that the synergies are recruited instantaneously without time delay for rapid grasps, and the synergy model from [31] was simplified as:

$$v(t) = \sum_{p=1}^{P} c_p S_p$$

where $v(t)$ represents the angular velocity of a joint at time t, c is the amplitude coefficient of synergy S, and P is the total number of synergies selected. The first six synergies were selected to reconstruct the hand kinematics here since over 80% variance was observed. A multivariate linear regression model was used to determine the relationship between neural features and the synergy coefficients:

$$C = X\beta$$

C represents the weights of synergies, determined by kinematic synergy-based reconstruction, and X is neural features. Decoding accuracy between synergy reconstructed kinematics and neural decoded kinematics were measured using a Pearson correlation coefficient.

3   Results

Experiment I: An average classification accuracy of 79 ± 5.5% (best up to 88 ± 6%) was achieved from all subjects on broad frequency range (1–45 Hz) in offline analysis with stacked autoencoders while average classification accuracies of 68 ± 9.1% (best up to 74 ± 9.1%) were achieved with PCA-LDA. To visualize the performance of the two classifiers (autoencoder and PCA-LDA), the receiver operating characteristic (ROC) curve and area under-curve (AUC) were applied, which are more representative of the performance of the classifiers than the classification accuracy rates, as shown in Fig. 2. Stacked autoencoders led to the best performance. Reaching direction is better classified from the 1–45 Hz frequency range, perhaps because more PSD features were included.

Experiment II: EEG features of hand grasp movement from the broad frequency range of 1–45 Hz were used to successfully decode synergy-based movements with an average decoding accuracy of 80.1 ± 6.1% (best up to 93.4 ± 2.3%). Figure 3 shows the angular velocity trajectories of recorded kinematics, synergy-based reconstructed kinematics, and neural decoded kinematics. Results showed that hand kinematics could be reconstructed accurately from synergies, and synergy weights could be successfully decoded from corresponding neural signals. For simple grasps, such as grasping a water bottle or petri dish, decoded kinematics are close to recoded kinematics, while a higher standard deviation was observed for precision grasp such as grasping a bracelet.

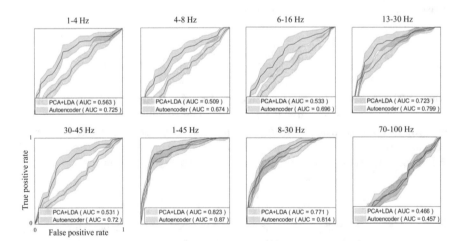

Fig. 2 The capacity of stacked autoencoders versus PCA-LDA from subject 4. With a similar classification accuracy rates observed from these two classifiers, stacked autoencoders performed better then PCA-LDA

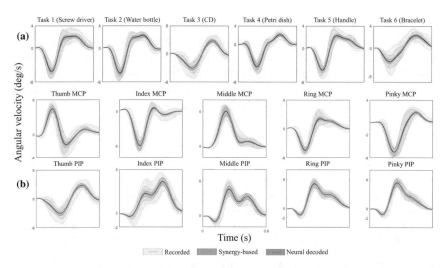

Fig. 3 Trajectory of hand kinematics. The recorded joint kinematics (green), reconstructed kinematics based on synergy (red) and neural decoded kinematics (blue) were perfectly matched. The shaded regions show standard deviations. **a** Kinematics of index MCP joints in each grasp type from Subject 2. **b** Kinematics of each joint in task 3 from Subject 2

4 Discussion, Conclusion and Future Directions

The arm reaching directions and hand grasp movements were successfully decoded from scalp EEG, and we demonstrated that more relevant information was encoded in broad frequency range EEG activity. In previous studies, neural activities in the low delta band (0–2 Hz) were widely used to decode reaching direction [14] and hand kinematics (velocity and position) [32]. However, it was pointed out that these studies may actually overestimate the decoding performance because of the natural relationship between slow cortical potentials and slow hand movements in the time domain [33]. This issue does not exist in the frequency domain. Additionally, spectral based models provide better performance in decoding hand motor function [34]. Our study included rapid grasps, rather than natural grasps, decoded from EEG spectrums (spectral powers in 1–45 Hz frequency range), thus avoiding overestimation.

Traditional EEG analysis approaches (such as PCA, LDA) dedicate great effort to linear neural decoding. Nevertheless, the complex nonlinear relationships embedded in noisy neural signals may not be fully represented by linear approaches. As machine leaning becomes increasingly popular, the algorithms become more useful in learning multiple levels of representation of both linear and nonlinear correlations. However, the small dataset in our experiment may restrict the wide utility of machine leaning algorithms, where this case can be easily attributed to the overfitting problem.

Reaching and grasping are fundamental actions in interacting with the world around us. Unfortunately, many individuals with hand movement disabilities lack these functions. Currently, lost limb functions are being restored using innovative invasive and noninvasive BCIs [35–37]. Improving motor control from high level synergies to lower level joints with noninvasive technologies remains a challenge. In this study, asynchronous arm reaching and synchronous hand grasping movements were decoded from broad range EEG activities in the sensorimotor area. With those outcomes, we envision using advanced algorithms to decode user intention, leading to real-time hand movement control in the near future.

References

1. S.R. Soekadar, N. Birbaumer, M.W. Slutzky, L.G. Cohen, Brain-machine interfaces in neurorehabilitation of stroke. Neurobiol. Dis. **83**, 172–179 (2015)
2. L. Randazzo, I. Iturrate, R. Chavarriaga, R. Leeb, J.D.R. Millan, Detecting intention to grasp during reaching movements from EEG. Proc. Annu. Int. Conf. IEEE Eng. Med. Biol. Soc. EMBS **2015**, 1115–1118 (2015)
3. N. Bernstein, *The co-ordination and regulation of movements.* Elsevier (1967)
4. M.C. Tresch, P. Saltiel, E. Bizzi, The construction of movement by the spinal cord. Nat. Neurosci. **2**(2), 162–167 (1999)
5. M. Santello, M. Flanders, J.F. Soechting, Postural hand synergies for tool use. J. Neurosci. **18**(23), 10105–10115 (1998)
6. S.M. Beurze, S. Van Pelt, W.P. Medendorp, Behavioral reference frames for planning human reaching movements. J. Neurophysiol. **96**(1), 352–362 (2006)
7. J. Fernandez-Ruiz, H.C. Goltz, J.F.X. DeSouza, T. Vilis, J.D. Crawford, Human parietal 'reach region' primarily encodes intrinsic visual direction, not extrinsic movement direction, in a visual-motor dissociation task. Cereb. Cortex **17**(10), 2283–2292 (2007)
8. Y.E. Cohen, R.A. Andersen, A common reference frame for movement plans in the posterior parietal cortex. Nat. Rev. Neurosci. **3**(7), 553–562 (2002)
9. E. Hoshi, J. Tanji, Integration of target and body-part information in the premotor cortex when planning action. Nature **408**(6811), 466–470 (2000)
10. E. Lew, Detection of self-paced reaching movement intention from EEG signals. Front. Neuroeng. **5**(July), 1–17 (2012)
11. O. Bai et al., Prediction of human voluntary movement before it occurs. Clin. Neurophysiol. **122**(2), 364–372 (2011)
12. E. Demandt, C. Mehring, K. Vogt, A. Schulze-Bonhage, A. Aertsen, T. Ball, Reaching movement onset- and end-related characteristics of EEG spectral power modulations. Front. Neurosci. **6**(MAY), 1–11 (2012)
13. T. Ball et al., Movement related activity in the high gamma range of the human EEG. Neuroimage **41**(2), 302–310 (2008)
14. E.Y.L. Lew, R. Chavarriaga, S. Silvoni, J.R. del Millán, Single trial prediction of self-paced reaching directions from EEG signals. Front. Neurosci. **8**(8), 1–13 (2014)
15. B. Libet, E.W. Wright, C.A. Gleason, Readiness-potentials preceding unrestricted 'spontaneous' versus pre-planned voluntary acts. Electroencephalogr. Clin. Neurophysiol. **54**(3), 322–335 (1982)
16. R. Scherer et al. *Brain-computer interfacing for assistive robotics*, vol. 14, July 2013
17. B. Awwad Shiekh Hasan, J.Q. Gan, Unsupervised movement onset detection from EEG recorded during self-paced real hand movement. Med. Biol. Eng. Comput. **48**(3), 245–253 (2010)
18. M. Santello, G. Baud-bovy, H. Jörntell, Neural bases of hand synergies. Front. Comput. Neurosci. **7**, 23 (2013)
19. M.T. Turvey, Action and perception at the level of synergies. Hum. Mov. Sci. **26**(4), 657–697 (2007)

20. R. Gentner, J. Classen, Modular organization of finger movements by the human central nervous system. Neuron **52**, 731–742 (2006)
21. M.C. Tresch, A. Jarc, The case for and against muscle synergies. Curr. Opin. Neurobiol. **19**(6), 601–607 (2009)
22. A.B. Schwartz, D.W. Moran, J. Jay, H. Drive, S. Diego, Arm trajectory and representation of movement processing in motor cortical activity. Eur. J. Neurosci. **12**, 1851–1856 (2000)
23. D.W. Moran, A.B. Schwartz, Motor cortical activity during drawing movements: population representation during spiral tracing. Am. Physiol. Soc. **82**(5), 2693–2704 (1999)
24. P. Braido, X. Zhang, Quantitative analysis of finger motion coordination in hand manipulative and gestic acts. Hum. Mov. Sci. **22**, 661–678 (2004)
25. R. Vinjamuri, M. Sun, C. Chang, H. Lee, R.J. Sclabassi, Z. Mao, Temporal postural synergies of the hand in rapid grasping tasks. IEEE Trans. Technol. Biomed. **14**(4), 986–994 (2010)
26. C. Kertzman, U. Schwarz, T.A. Zeffiro, M. Hallett, The role of posterior parietal cortex in visually guided reaching movements in humans. Exp. Brain Res. **114**(1), 170–183 (1997)
27. G. Schalk, D.J. McFarland, T. Hinterberger, N. Birbaumer, J.R. Wolpaw, BCI2000: a general-purpose brain-computer interface (BCI) system. IEEE Trans. Biomed. Eng. **51**(6), 1034–1043 (2004)
28. D. Pei, M. Burns, R. Chandramouli, R. Vinjamuri, Decoding asynchronous reaching in electroencephalography using stacked autoencoders. IEEE Access **6**, 52889–52898 (2018)
29. V. Patel, M. Burns, D. Pei, R. Vinjamuri, Decoding synergy-based hand movements using electroencephalography, in *2018 40th Annual International Conference on IEEE Engineering Medicine Biology Society* pp. 4816–4819 (2018)
30. D. Pei, S. Member, V. Patel, Neural decoding of synergy-based hand movements using electroencephalography. IEEE Access **7**, 18155–18163 (2019)
31. V. Ramana, M. Sun, C. Chang, H. Lee, R.J. Sclabassi, Dimensionality reduction in control and coordination of the human hand. IEEE Trans. Biomed. Eng. **57**(2), 284–295 (2010)
32. H. Agashe, J.L. Contreras-Vidal, Reconstructing hand kinematics during reach to grasp movements from electroencephalographic (EEG) signals, in *33rd Annual International Conference of the IEEE EMBS*, pp. 5444–5447 (2011)
33. J.M. Antelis, L. Montesano, A. Ramos-murguialday, N. Birbaumer, J. Minguez, On the usage of linear regression models to reconstruct limb kinematics from low frequency EEG signals. PLoS ONE **8**(4), e61976 (2013)
34. A. Korik, R. Sosnik, N. Siddique, D. Coyle, Decoding imagined 3D hand movement trajectories from EEG: evidence to support the use of mu, beta, and low gamma oscillations. Front. Neurosci. **12**(March), 1–16 (2018)
35. J.L. Collinger et al., High-performance neuroprosthetic control by an individual with tetraplegia. Lancet **381**(9866), 557–564 (2013)
36. S. Qiu, Z. Li, W. He, L. Zhang, Brain—machine interface and visual compressive sensing-based teleoperation control of an exoskeleton robot. IEEE Trans. Fuzzy Syst. **25**(1), 58–69 (2017)
37. W. Wang et al., An electrocorticographic brain interface in an individual with tetraplegia. PLoS ONE **8**(2), e55344 (2013)

Restoring Functional Reach-to-Grasp in a Person with Chronic Tetraplegia Using Implanted Functional Electrical Stimulation and Intracortical Brain-Computer Interfaces

A. Bolu Ajiboye, Leigh R. Hochberg, and Robert F. Kirsch

Abstract This study demonstrates volitional arm and hand motions restored to a person living with complete tetraplegia due to high cervical spinal cord injury. Selective intramuscular functional electrical stimulation (FES) of paralyzed muscles throughout the upper extremity powered multiple reaching and grasping movements. An intracortical brain computer interface (iBCI) recorded neural signals from the participant's contralateral motor cortex, extracted movement intentions from these signals, and commanded FES patterns to generate these intended movements. As a result of the combined technological approach, the participant could volitionally reach, grasp, and drink from a cup, demonstrating the feasibility of this FES + iBCI system to restore cortically-controlled functional arm and hand movements in persons with extensive paralysis.

Keywords Spinal cord injury (SCI) · Intracortical brain computer interface (iBCI) · Functional electrical stimulation (FES) · Tetraplegia · Microelectrode arrays

A. Bolu Ajiboye (✉) · R. F. Kirsch
Department of Biomedical Engineering, Case Western Reserve University, 10900 Euclid Ave, Cleveland, OH 44106, USA
e-mail: aba20@case.edu

Department of Veterans Affairs Medical Center, Louis Stokes Cleveland, FES Center of Excellence, Rehab. R&D Service, 10701 East Blvd, Cleveland, OH 44106, USA

School of Medicine, Case Western Reserve University, 10900 Euclid Ave, Cleveland, OH 44106, USA

L. R. Hochberg
School of Engineering, Brown University, Providence, RI, USA

Rehabilitation R&D Service, Department of Veterans Affairs Medical Center, Center for Neurorestoration and Neurotechnology, Providence, RI, USA

Brown Institute for Brain Science, Brown University, Providence, RI, USA

Department of Neurology, Massachusetts General Hospital, Boston, MA, USA

Department of Neurology, Harvard Medical School, Boston, MA, USA

© The Author(s), under exclusive license to Springer Nature Switzerland AG 2020
C. Guger et al. (eds.), *Brain–Computer Interface Research*,
SpringerBriefs in Electrical and Computer Engineering,
https://doi.org/10.1007/978-3-030-49583-1_4

1 Clinical Significance and Background

Spinal cord injury (SCI) resulting in paralysis affects over 250,000 people nationwide with over 12,000 new cases each year. Slightly more than half of all SCI cases occur at cervical levels (tetraplegia). Incomplete and complete tetraplegia have accounted respectively for 34 and 18% of all SCI cases since 2000, with less than 1% of all cases achieving full recovery [1]. Even with a caregiver, many of these individuals experience a lower quality of life due to loss of personal independence from the inability to perform standard activities-of-daily-living (ADL) on their own, such as reaching to grasp objects for drinking and self-feeding. People with low cervical SCI (C5-C7) resulting in chronic hand paralysis can regain assisted arm reaching and simple hand grasp function by using functional electrical stimulation (FES) neuroprostheses, using their residual voluntary movement to command the neuro-prosthesis. FES, in the absence of descending cortical commands, applies spatially and temporally coordinated patterns of electrical stimulation to peripheral nerves and muscles to reanimate paralyzed limbs and allow for performance of simple but functional and meaningful arm and hand movements [2–5]. In particular, FES has been used to restore simple arm and hand function to persons with cervical SCI using position transducer and muscle (electromyogram, or EMG) command interfaces [2, 4, 6]. In contrast, people with high cervical SCI (C1-C4), resulting in chronic tetraplegia, have limited residual voluntary movement post-injury to use as a command source, and hence cannot adequately command even simple FES restored movements without implementing a more suitable command interface, such as a brain interface.

Recent investigations have shown the efficacy of using intracortical brain-computer interfaces (iBCIs) to decipher intended movement, from the electrical activity of intact cortical networks, to command various simple external devices. The possibility of naturally commanding neuroprosthetic devices, such as FES arm and hand systems, using an iBCI offers significant potential benefit to people with chronic and complete tetraplegia. Current rehabilitation command options for people with high cervical SCI have historically included mouthsticks, chin-controlled joysticks, sip-and-puff [7], and voice recognition systems. However, these systems are slow and limited in functionality, which makes them unsuitable for commanding neuro-prostheses to perform multi-dimensional coordinated and dexterous actions. An iBCI commanded neuroprosthesis offers the possibility of harnessing intact cortical activities that remain even many years after injury. Patterns of cortical activation related to reaching and grasping have been validated in both non-human primate models [8, 9] and human participants. By recording these cortical patterns, extracted intended movement signals have been used to command higher dimensional prosthetic systems [10, 11]. Persons with paralysis are very knowledgeable of, and highly amenable to, receiving iBCIs for commanding neuroprosthetic upper extremity movements, provided that there are substantial performance gains over less invasive options [12, 13]. Hence iBCIs, when combined with FES, may offer a user an acceptable means of providing advanced functional arm and hand movement restoration. Persons with arm and hand paralysis have stated that they would much prefer to regain command over their own limbs (reanimated via FES) than other types of movement assistive devices, such as robotic limbs [12, 14]. There is a potential psychological benefit to seeing one's own limb move [15], possibly contributing to this preference. Furthermore, robotic assistants have met with limited acceptance in the past [16, 17] because of the inconvenience of setup and limited portability [18].

Our team of investigators at Case Western Reserve University demonstrated that a person with chronic and complete tetraplegia could use an invasive BCI to command both reaching and grasp movements of his paralyzed arm, restored by FES [19]. The study, described in detail in the remainder of this chapter, used intramuscularly implanted FES electrodes in combination with dual intracortical microelectrodes and a cortically commanded motorized arm support to restore movements of the shoulder, elbow, wrist, and hand grasping. We show that, using this combined FES + iBCI technology, the study participant regained the ability to perform functional and meaningful reaching and grasping movements. This study is the first demonstration of an implantable BCI with an implanted stimulation system, and shows the potential for a totally implanted BCI and FES system that would restore cortically commanded whole arm and grasping movements to persons with paralysis.

2 System Description and Scientific Approach

The study participant (identified as T8) was enrolled into the BrainGate2 pilot clinical trial (ClinicalTrials.gov NCT00912041) and gave informed consent for medical and research procedures as approved by the Institutional Review Boards of University Hospitals Case Medical Center (Cleveland, OH) and Massachusetts General Hospital (Boston, MA). At the time of this study, T8 was a 53-year-old man with high cervical SCI (C4, AIS A) that occurred 8 years prior to enrollment. On his right side (contralateral to the intracortical implant), T8 retained some limited voluntary shoulder girdle motion, but no voluntary glenohumeral, elbow, or hand function. T8 underwent three separate surgical procedures. First, he received two 96-channel microelectrode arrays (Blackrock Microsystems, Salt Lake City, Utah) [20] that were implanted into the hand area on the precentral gyrus [21] of his motor cortex (Fig. 1b). During two subsequent procedures, occurring respectively four and nine months post implantation of the arrays, T8 received a total of thirty-six percutaneous muscle stimulating electrodes (Synapse Biomedical, Oberlin, OH) [4] (Fig. 1c) in his right (dominant) upper and lower arm. These included four percutaneous anodic current return electrodes. Implanted muscles include those for finger (flexor digitorum superficialis, extensor digitorum communis), thumb (flexor pollicis longus, adductor pollicis, extensor pollicis longus), wrist (extensor carpi radialis/ulnaris, flexor carpi radialis/ulnaris), elbow (biceps, triceps), and shoulder (anterior, posterior deltoids, pectoralis major) functions. By stimulating the thumb and finger muscles, we could restore a lateral hand grasp (where the thumb pad opposed the lateral surface

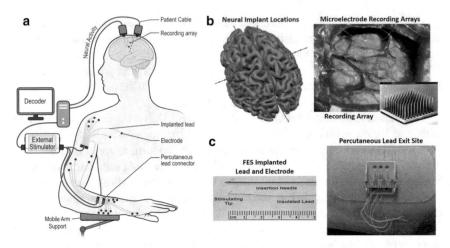

Fig. 1 Overview of the FES + iBCI system. **a** Illustration of how the system components are connected, **b** [left] structural MRI scan of T8's brain. The implant locations of the two intracortical arrays are indicated with red squares. [top right] Photo of microelectrode arrays and wire bundles shortly after intracortical implantation. [bottom right] SEM of an example microelectrode recording array (photo courtesy of Blackrock Microsystems), **c** example FES Implanted lead and electrode

of the proximal phalanx of the index finger), which can be used to complete a wide range of functional tasks [22, 23]. All implanted muscles were exercised 2–3 times per week, 2–4 h per session using cyclical electrical stimulation patterns to improve strength, range of motion, and fatigue resistance. Electrical stimulation resulted in restoration of 76 degrees of flexion/extension of the elbow, and a lateral grasp that could close with enough force to securely hold several objects, including a coffee mug.

Figure 1a illustrates the percutaneous FES + iBCI system. The implanted microelectrode arrays recorded electrical activity from multiple neurons. A neural decoder then translated the recorded neural activity patterns into command signals for controlling the stimulation of biceps, triceps, forearm, and hand muscles to produce coordinated reaching and grasping movements. An external stimulator delivered charge-balanced, biphasic, constant-current stimulation through the percutaneous muscle stimulation electrodes, with fixed current amplitude (20 mA) and frequency (12.5 Hz), and variable pulse duration of 0–200 μs, to produce muscle contractions and subsequent limb movement. The duration of the current pulse ("pulse-width") applied at a given electrode determined the strength of the muscle contraction, and different muscles were activated in a coordinate manner to produce functional arm and hand movements. Thus, we varied the pulse-widths and muscle combinations as functions of T8's cortically-derived movement commands to enable graded control of functional motions. To support the paralyzed arm against gravity during movement, T8 was fitted with a Mobile Arm Support (MAS) (Focal MEDITECH, Tilburg, Netherlands). The MAS also provided a motorized vertical arm motion that was used to restore self-initiated, cortically-controlled shoulder elevation. Instrumented goniometers were fitted onto T8's elbow, wrist, and/or hand to measure the range of motion of each actuated movement.

A neural decoder was calibrated at the beginning of each experimental session to translate patterns of cortical neural activity into command signals for the FES system. The decoder used two neural features from each electrode of the intracortical microelectrode arrays: (1) the threshold crossing (TX) "firing" rate, determined by counting the number of action potentials present in a 20 ms time window that crossed a preset noise threshold, and (2) the average spectral high frequency power (HFP, 250–3000 Hz) in a 20 ms time window. The decoder used a linear transformation function, similar to the Kalman filter used in recent iBCI applications [24], to map the 384 features (192 firing rates and 192 spectral high frequency power values) to three movement commands. Each command determined the stimulation level corresponding to a specific pattern of muscles (either a hand open/close pattern or an elbow flexion/extension pattern) or the actuation level of the mobile arm support. Figure 2a illustrates the decoding process.

To initialize the neural decoder, we used neural data recorded while T8's arm was automatically driven by the FES system to make elbow, hand or shoulder movements, and he was verbally instructed to simultaneously attempt to control the observed arm motions. We then refined the decoder [25] by using neural data recorded while T8 actually controlled his FES-powered arm movements, using the initial decoder to volitionally make those same movements. After calibration and refinement, the neural decoder was held constant for the remainder of the research session while T8 made several single joint movements and completed a functional coffee drinking task.

3   Results

Neural activity recorded from many of the intracortical electrodes was strongly related to T8's intended movement commands when he attempted to use the FES + iBCI system. Figure 2b illustrates how the threshold crossing rates observed on two example electrodes changed as a function of the movement that T8 was attempting in response to verbal instruction. On one electrode, substantially more threshold crossings were observed during attempted elbow flexion as opposed to extension, and on another electrode, substantially more threshold crossings were observed during attempted hand opening as opposed to closing. Of the 192 electrodes, we identified a neural feature (either threshold crossing or spectral high frequency power) that coded for hand opening and closing on 36 electrodes, for elbow flexion and extension on 45 electrodes, and for mobile arm support elevation on 37 electrodes. We considered a neural feature to "code" for a certain movement if that feature's mean value was significantly different (t-test, $p < 10^{-4}$) between the two opposing commands, such as hand opening versus closing.

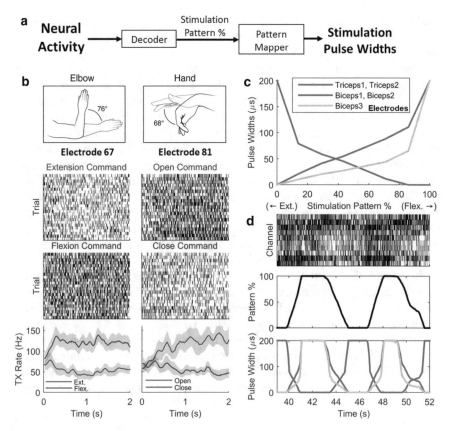

Fig. 2 Overview of how the FES + iBCI system translates cortical neural activity into FES stimulus parameters. **a** Neural activity is decoded into a single command signal for each joint ("stimulation pattern %"). A pattern mapper then converts this command signal into the appropriate pulse widths to apply to each individual FES electrode, enabling the participant to coordinate the action of multiple electrodes and muscles using only a single command. **b** Example neural activity during elbow flexion/extension commands (electrode 67) and hand opening/closing commands (electrode 81). **c** The stimulation pattern used for converting a decoded elbow movement command into the stimulation pulse widths required to produce the commanded movement. **d** Example threshold crossing rates, decoded command, and pulse widths over time as T8 made elbow flexion and extension movements

T8 completed a series of self-initiated, iBCI-commanded FES arm reaching movements involving 2D (elbow, grasp) and 3D (elbow, shoulder, grasp) joint movements. Sitting upright in his wheelchair, T8 first qualitatively demonstrated the ability to command movement of each degree-of-freedom independent. The first successful demonstration of T8 cortically commanding movements of his reanimated arm and hand occurred 7 days post-implant of the FES electrodes. After demonstrating robust control of single degree-of-freedom arm and hand movements, T8 then demonstrated his ability to perform an activity-of-daily living (ADL) task in which he acquired a cup of coffee and took a drink (Fig. 3A). The task required T8 to use the FES + iBCI system to (1) extend his elbow, (2) open his hand, (3) grasp the cup securely, (4) flex his elbow to transport it close to his mouth, 5) take a drink, 6) extend his elbow to return the cup, and 7) release his grasp. T8 required between 20 to 40 s to complete the drinking task and was successful in 11 of the 12 attempts made during the illustrated session (Fig. 3b). His success demonstrates for the first time that a person with extensive paralysis, and nine years post injury, can perform cortically-controlled volitional functional movements involving both arm reaching and hand grasping with an FES + iBCI system. When asked to describe how he commanded the FES arm movements, T8 replied, "It's probably a good thing that I'm making

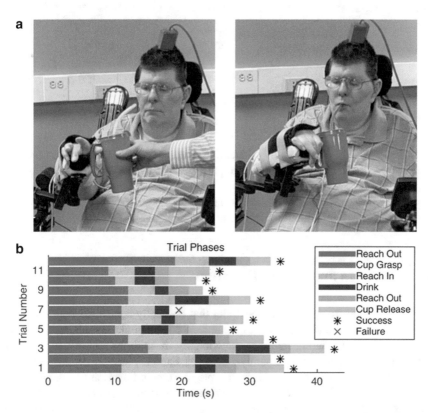

Fig. 3 T8 using the FES + iBCI system to take a drink of coffee. **a** T8 reaching out to grasp the cup of coffee (left) and bringing it to his mouth to take a drink (right). **b** The length of time it took T8 to complete each phase of the drinking task. Data is shown for 12 trials completed within a single experimental session; only one trial was failed when T8 dropped the cup. **c** Example time series of T8's elbow and hand motion when the FES + iBCI system was turned on (left) and when the FES system was turned off (right). When the system was on, the decoded neural commands (blue) and the elbow and hand joint angles (orange) changed appropriately as T8 moved through the phases of the task, enabling him to take a drink of coffee. When the system was off, T8 could only make small, uncontrolled elbow jerks caused by his residual shoulder motion and could not move his hand at all

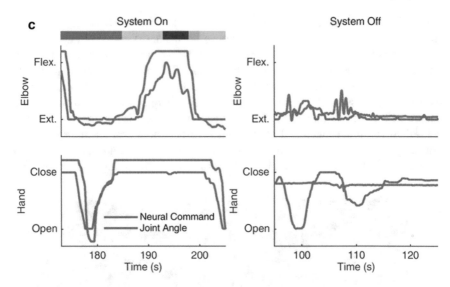

Fig. 3 (continued)

it move without having to really concentrate hard at it....I just think 'out' and... it just goes." T8 was completely unable to perform the same task with the FES system turned off (Fig. 3c); his minor residual motion of his shoulder girdle could only cause a small, uncontrolled jerk of his elbow and could not move his hand at all, despite T8's attempts to command the required arm movements.

4   Discussion and Future Direction

We have demonstrated, for the first time, that cortically-controlled volitional arm and hand function can be restored to a person with chronic tetraplegia by simultaneously (1) reanimating multiple, functionally meaningful motions of the limb through intramuscular FES of paralyzed muscles and (2) enabling control of these FES-restored motions by extracting multiple movement intention commands from intracortical recordings in real time. The simultaneous implementation of iBCI and FES technologies represents a neurotechnology-based bridging of the participant's spinal cord injury and demonstrates a significantly more intuitive command interface than those currently available to persons with extensive (whole arm) paralysis. With the FES + iBCI system, T8 was once again able to just "think" about moving his arm and hand, and the movement intentions decoded from the recorded neural activity were sufficient to create the desired arm and hand movements via the FES system.

By restoring iBCI-commanded and FES-driven motion of both the arm and hand in a human participant, the present work significantly extends previous iBCI research performed in intact [26–28] and temporarily paralyzed non-human-primates [8, 9], as well as research done in individuals with paralysis controlling cursors or robotic limbs [10, 11, 29, 30]. Our present FES + iBCI system restores both reaching and grasping to an individual with complete motor paralysis of the hand and arm, while following a viable path to clinical relevance through the use of implantable stimulation technology.

The performance of the FES + iBCI system used here was somewhat limited by SCI-related conditions such as muscle atrophy, denervation, and joint contractures, and by the difficulty of precisely targeting desired muscles when inserting FES electrodes percutaneously rather than in an open surgical procedure. However, these limitations are addressable by currently available, implantable FES technologies (e.g., precisely located intramuscular electrodes and peripheral nerve cuff electrodes) and associated techniques (model-based optimization of muscle stimulation patterns, muscle tendon transfers to replace the functions of denervated muscles, and more extensive exercise programs). Advances in intracortical electrodes to enhance long-term recording stability (through increased mechanical and biological viability [31, 32]), and to enable a fully-implanted brain recording interface [33] may also increase the clinical viability of an FES + iBCI system. Nevertheless, (1) T8 had a high rate of success in performing a reaching and grasping task using his current percutaneous FES-activated arm and hand, and (2) remaining performance limitations can be largely addressed using existing (but permanent) technologies. The movements afforded to T8 by the current system allowed him to take a drink of coffee, with his own arm and hand, solely of his own volition (reaching out, grasping, reaching back to the face). These actions are representative of movements needed to perform a wide range of reaching tasks [34], suggesting that more functional activities may be achieved with the current system. The present study used percutaneous, removable technologies both for the brain recordings and the FES system in order to first evaluate the feasibility of the approach. Future systems inspired by this project are being designed to provide full-time, reliable, effective, intuitive control of the arm and hand, and may thus enable restoration of a much wider range of functional activities.

Acknowledgements We thank the study participant for his pioneering efforts participating in the present study. We thank the Louis Stokes Cleveland VA Medical Center Cares Tower Residence Center, for space and logistical support. We additionally thank the members of the BrainGate2 Consortium for their feedback and support of the research efforts.

Support for this work was provided by the National Institutes of Health under grants NIH 1R01HD077220, NIH N01HD53403, NIH R01DC009899, and VA B6453R. The reported contents do not necessarily represent the views of the funding or parent institutions, or of the US Government. The funding sources had no role in the writing of this manuscript.

References

1. NSCISC, Spinal cord injury, facts and figures at a glance. J. Spinal Cord. Med. **36**, 1–2 (2013)
2. K.L. Kilgore, H.A. Hoyen, A.M. Bryden, R.L. Hart, M.W. Keith, P.H. Peckham, An implanted upper-extremity neuroprosthesis using myoelectric control. J. Hand. Surg. [Am.] **33**, 539–550 (2008)
3. P.H. Peckham, M.W. Keith, K.L. Kilgore et al., Efficacy of an implanted neuroprosthesis for restoring hand grasp in tetraplegia: a multicenter study. Arch. Phys. Med. Rehabil. **82**, 1380–1388 (2001)
4. W.D. Memberg, K.H. Polasek, R.L. Hart et al., Implanted neuroprosthesis for restoring arm and hand function in people with high level tetraplegia. Arch. Phys. Med. Rehabil. **95**, 1201–1211 (2014)
5. P.H. Peckham, J.S. Knutson, Functional electrical stimulation for neuromuscular applications. Ann. Rev. Biomed. Eng. **7**, 327–360 (2005)
6. Kirsch RF (2008) Restoration of hand and arm by functional neuromuscular stimulation. In: Neural Interfaces Conference. Cleveland, OH, 2008
7. P.A. Lathem, T.L. Gregorio, S.L. Garber, High level quadriplegia: an occupational therapy challenge. Am. J. Occup. Ther. **39**, 705–714 (1985)
8. C. Ethier, E.R. Oby, M.J. Bauman, L.E. Miller, Restoration of grasp following paralysis through brain-controlled stimulation of muscles. Nature **485**, 368–371 (2012)
9. C.T. Moritz, S.I. Perlmutter, E.E. Fetz, Direct control of paralysed muscles by cortical neurons. Nature **456**, 639–642 (2008)
10. L.R. Hochberg, M.D. Serruya, G.M. Friehs et al., Neuronal ensemble control of prosthetic devices by a human with tetraplegia. Nature **442**, 164–171 (2006)
11. L.R. Hochberg, D. Bacher, B. Jarosiewicz et al., Reach and grasp by people with tetraplegia using a neurally controlled robotic arm. Nature **485**, 372–375 (2012)
12. C.H. Blabe, V. Gilja, C.A. Chestek, K.V. Shenoy, K.D. Anderson, J.M. Henderson, Assessment of brain-machine interfaces from the perspective of people with paralysis. J. Neural Eng. **12**, 043002 (2015)
13. J. Lahr, C. Schwartz, B. Heimbach, A. Aertsen, J. Rickert, T. Ball, Invasive brain–machine interfaces: a survey of paralyzed patients' attitudes, knowledge and methods of information retrieval. J. Neural Eng. **12**, 043001 (2015)
14. J.L. Collinger, M.L. Boninger, T. Bruns, K. Curley, W. Wang, D.J. Weber, Functional priorities, assistive technology, and brain-computer interfaces after spinal cord injury. J. Rehabil. Res. Dev. **50**, 145–160 (2013)
15. R.H. Nathan, A. Ohry, Upper limb functions regained in quadriplegia: a hybrid computerized neuromuscular stimulation system. Arch. Phys. Med. Rehabil. **71**, 415–421 (1990)
16. J. Hammel, K. Hall, D. Lees et al., Clinical evaluation of a desktop robotic assistant. J. Rehabil. Res. Dev. **26**, 1–16 (1989)
17. J.M. Hammel, H.F. Van der Loos, I. Perkash, Evaluation of a vocational robot with a quadriplegic employee. Arch. Phys. Med. Rehabil. **73**, 683–693 (1992)
18. R. Mahoney, Robotic products for rehabilitation: status and strategy. In: International Conference on Rehabilitation Robotics. Bath, UK, 1997: 1–6
19. A.B. Ajiboye, F.R. Willett, D.R. Young et al., Restoration of reaching and grasping movements through brain-controlled muscle stimulation in a person with tetraplegia: a proof-of-concept demonstration. Lancet **6736**, 1–10 (2017)
20. E.M. Maynard, C.T. Nordhausen, R.A. Normann, The Utah intracortical electrode array: a recording structure for potential brain-computer interfaces. Electroencephalogr. Clin. Neurophysiol. **102**, 228–239 (1997)
21. T.A. Yousry, U.D. Schmid, H. Alkadhi et al., Localization of the motor hand area to a knob on the precentral gyrus. A New Landmark. Brain **120**, 141–157 (1997)
22. K.L. Kilgore, P.H. Peckham, G.B. Thrope, M.W. Keith, K.A. Gallaher-Stone, Synthesis of hand grasp using functional neuromuscular stimulation. IEEE Trans. Biomed. Eng. **36**, 761–770 (1989)

23. University of California LAD of E. Studies to Determine the Functional Requirements for Hand and Arm Prosthesis: The Final Report Covering Work During the Year 1946–1947, Under Subcontract No. 17 of Prime Contract VAm-21223 with the National Academy of Sciences. 1947

24. D. Bacher, B. Jarosiewicz, N.Y. Masse, et al., Neural point-and-click communication by a person with incomplete locked-in syndrome. *Neurorehabil Neural Repair* 2014; published online Nov. https://doi.org/10.1177/1545968314554624

25. B. Jarosiewicz, N.Y. Masse, D. Bacher et al., Advantages of closed-loop calibration in intra-cortical brain—computer interfaces for people with tetraplegia. J. Neural Eng. **10**, 046012 (2013)

26. D.M. Taylor, S.I. Helms-Tillery, A.B. Schwartz, Direct cortical control of 3D neuroprosthetic devices. Science (80–) **296**, 1829–1832 (2002)

27. M.D. Serruya, N.G. Hatsopoulos, L. Paninski, M.R. Fellows, J.P. Donoghue, Instant neural control of a movement signal. Nature **416**, 141–142 (2002)

28. M. Velliste, S. Perel, M.C. Spalding, A.S. Whitford, A.B. Schwartz, Cortical control of a prosthetic arm for self-feeding. Nature **453**, 1098–1100 (2008)

29. J.L. Collinger, B. Wodlinger, J.E. Downey et al., High-performance neuroprosthetic control by an individual with tetraplegia. Lancet **381**, 557–564 (2013)

30. T.N. Aflalo, S. Kellis, C. Klaes et al., Decoding motor imagery from the posterior parietal cortex of a tetraplegic human. Sci Mag **348**, 906–910 (2015)

31. M. Jorfi, J.L. Skousen, C. Weder, J.R. Capadona, Progress towards biocompatible intracortical microelectrodes for neural interfacing applications. J. Neural Eng. **12**, 011001 (2015)

32. J.C. Barrese, N. Rao, K. Paroo et al., Failure mode analysis of silicon-based intracortical microelectrode arrays in non-human primates. J. Neural Eng. **10**, 066014 (2013)

33. D. Borton, M. Yin, J. Aceros, A. Nurmikko, An implantable wireless neural interface for recording cortical circuit dynamics in moving primates. J. Neural Eng. **10**, 026010 (2013)

34. A.S. Cornwell, J.Y. Liao, A.M. Bryden, R.F. Kirsch, Standard task set for evaluating reha-bilitation interventions for individuals with arm paralysis. J. Rehabil. Res. Dev. **49**, 395 (2012)

Towards Speech Synthesis from Intracranial Signals

Christian Herff, Lorenz Diener, Emily Mugler, Marc Slutzky, Dean Krusienski, and Tanja Schultz

Abstract Brain-computer interfaces (BCIs) are envisioned to enable individuals with severe disabilities to regain the ability to communicate. Early BCIs have provided users with the ability to type messages one letter at a time, providing an important, but slow, means of communication for locked-in patients. However, natural speech contains substantially more information than a textual representation and can convey many important markers of human communication in addition to the sequence of words. A BCI that directly synthesizes speech from neural signals could harness this full expressive power of speech. In this study with motor-intact patients undergoing glioma removal, we demonstrate that high-quality audio signals can be synthesized from intracranial signals using a method from the speech synthesis community called Unit Selection. The Unit Selection approach concatenates speech units of the user to form new audio output and thereby produces natural speech in the user's own voice.

Keywords Brain-to-speech · ECoG · Speech · Brain-computer interface (BCI)

C. Herff (✉)
School for Mental Health and Neuroscience, Maastricht University, Maastricht, The Netherlands
e-mail: c.herff@maastrichtuniversity.nl

C. Herff · L. Diener · T. Schultz
Cognitive Systems Lab, University of Bremen, Bremen, Germany

E. Mugler · M. Slutzky
Department of Neurology, Physiology, and Physical Medicine & Rehabilitation, Northwestern University, Chicago, IL, USA

D. Krusienski
Biomedical Engineering Department, Virginia Commonwealth University (VCU), Richmond, VA, USA

© The Author(s), under exclusive license to Springer Nature Switzerland AG 2020
C. Guger et al. (eds.), *Brain–Computer Interface Research*,
SpringerBriefs in Electrical and Computer Engineering,
https://doi.org/10.1007/978-3-030-49583-1_5

1 Introduction

Brain-computer interface (BCI) research has made notable progress in recent years in enabling communication and control for healthy and impaired users [1]. Recent advances in the decoding of aspects of speech from neural signals, such as articulatory gestures [2], articulator kinematics [3], segmental features [4], phonemes [5, 6], key words [7] and continuous speech [8–10], might enable natural composing of messages in the future [11]. See [12] for a recent review of speech decoding efforts for brain-computer interfacing. To give BCI users the full expressive power of speech, the classification of a predefined set of classes (e.g. words) is not sufficient, as aspects such as prosody and accentuation are lost, which are necessary to convey emotion. Decoded words could be audibly synthesized by a text-to-speech (TTS) engine, but this would inherently introduce a delay of at least the length of the words, resulting in severe speech disruptions [13]. An approach that directly translates measured brain activity into audio could mitigate these problems and give the full expressive power of speech to BCI users.

Direct synthesis from neural recordings was first investigated by Guenther et al., who demonstrated real-time synthesis of vowel formants from intracortical spikes using neurotrophic electrodes in a paralyzed patient [14]. Pasley et al. extended these results to the synthesis of perceived speech from recordings in the auditory cortex [15]. Martin et al. further extended these results to the decoding of spatio-temporal features of speech production from intracranial recordings [16], but did not synthesize audio waveforms from these features. The idea behind these approaches is to reconstruct the speech spectrogram from the neural recordings and then create the audio waveform from the spectrogram, even though the phase information is lost. Figure 1 highlights the basic idea of this approach.

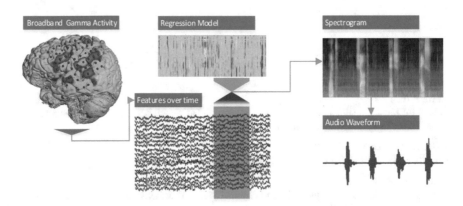

Fig. 1 Traditional approach in the conversion of neural activity to audio waveform: a regression model is used to map neural activity to a spectrogram, which is then transferred to an audio waveform

The regression model mapping neural activity to a spectrogram can be based on simple linear approaches [17] or even neural networks [18–20]. Another approach used articulatory kinematics as an intermediate step in the conversion [21].

Here, we present a relatively straightforward alternative approach that uses pattern matching to create natural sounding speech output by directly combining units of previously recorded speech. The reconstructed output can sound very natural because units of the user's own voice are concatenated. Our approach is based only on the measured neural activity and is fast enough for real-time processing. The presented approach might therefore enable natural conversation for BCI users in the future.

2   Methods

2.1   Experiment Design and Data Recording

For our study [22], parallel audio and neural activity recordings were obtained during surgery for brain tumor resection. During these surgeries, the eloquent cortex needs to be mapped to ensure that no part of the cortex is removed that is critical for speech or language function. We recorded ECoG during the surgery using an 8×8 array of electrodes (4-mm interelectrode spacing, 2.3-mm diameter contact size) placed over inferior frontal, premotor and motor cortices. After electrical stimulation mapping was finalized, participants voluntarily participated in our experiment. All participants gave written informed consent prior to the surgery, and the study was approved by the Institutional Review Board of Northwestern University.

During the experiment, participants were asked to read single words aloud that were shown to them on a computer screen. We analyzed data from 6 participants (one female) reading between 244 and 372 words (<12 min, 40 s). These data were part of a dataset used previously to investigate representation of articulatory gestures in motor, pre-motor and inferior frontal cortices [2].

2.2   Data Processing

To extract meaningful information from the ECoG signals, we extracted logarithmic high-gamma (70–170 Hz, HG) power in windows of 400 ms length (downsampled to 20 Hz) to capture the spatio-temporal dynamics of speech production in the cortex [23]. We calculate the logarithm of the HG power to make the distribution more Gaussian [24]. As acoustic properties change very quickly in continuous speech, we applied a frameshift of 10 ms, i.e. a new 400 ms window of logarithmic HG power is extracted every 10 ms. To maintain the alignment between audio and HG features, we dissected the audio data into 150 ms long units of raw audio data with a frameshift of 10 ms corresponding to the frameshift in the HG features.

2.3   Decoding Approach

For the generation of natural speech directly from neural signals, we used a method from the speech synthesis community called Unit Selection [25]. The Unit Selection approach is known to perform especially well when little data is available, and is therefore a good fit for our experiments with the extremely limited data set sizes. Unit Selection was originally used to create natural speech from text and was later used to map one voice to another (voice conversion). The idea behind the approach is to find the best fitting unit of speech given the input (new letter in TTS, or new unit of speech that should be converted) and concatenate it to the previous output. To find the best-fitting unit of speech, two cost terms are combined. The concatenation cost measures how well the candidate unit fits the previous unit, while the target cost measures how well the candidate unit fits the input.

We modified the Unit Selection approach to generate natural speech from HG features in the following manner. For each window of HG activity in the test data, we found the most similar unit of HG activity in the training data. The corresponding speech unit to this window HG activity was then concatenated to the previous output. We used the cosine similarity measure to establish similarity between windows of HG activity. In our approach, we only utilized the target cost, as including concatenation cost increases the computational cost of the approach.

Given the frameshift of 10 ms and the 150 ms long units of speech, a very large overlap between units of speech exists. This overlap is used to create very smooth output by combining the units of speech with Hamming windows. Our approach directly outputs audio waveforms, no intermediate representation of speech is necessary.

2.4   Evaluation

We evaluated our approach in a 5-fold cross-validation, in which 80% of the data were used for training and 20% of the data were used for testing in a round-robin manner until all data were used for testing exactly once. We made sure that no word was in both training and testing sets in each fold.

Raw audio waveforms can be difficult to compare, so we transferred the waveforms into a spectral representation that better represents human perception of speech by transforming it onto the mel-scale [26] using triangular filter banks. As a quantitative measure of reconstruction quality, we calculated the Pearson correlation between the original and reconstructed logarithmic mel-scaled spectrograms for each frequency bin individually.

To establish a chance level for our approach, we generated a randomized baseline by choosing a random unit of speech instead of the best fitting one. We then concatenated the units in the same way we did for the real reconstruction. This randomization

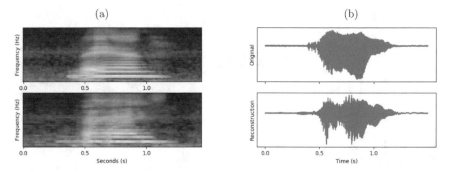

Fig. 2 **Reconstruction examples**: original (top) and reconstruction (bottom) of the word 'pace' in the spectral (**a**) and temporal (**b**) domain

was repeated 1000 times, and the highest resulting correlation was used to establish the chance-level baseline.

3   Results

The presented approach concatenates units of original speech by the user, which results in natural output that preserves many of the characteristics of natural speech. Figure 2 shows an example of a reconstructed word in time-domain and spectrogram. Especially in the spectral domain, the similarity is quite striking.

Inspecting correlations between original and reconstructed spectrograms, our approach achieved correlations significantly above chance level (best participant $r = 0.57$, mean $r = 0.25$, highest randomized $r = 0.03$). These correlation coefficients are consistent across all logarithmic mel-scaled coefficients, highlighting that all relevant speech information can be preserved (Fig. 3).

4   Conclusion

The high-quality reconstructions generated by our approach, many of which are intelligible to human listeners, show that a simple pattern matching approach such as Unit Selection can be used to generate speech without the need for complex machine learning models with thousands of trainable parameters. Our approach achieves good correlations despite not being trained to maximize correlations, as the approach does not operate in the spectral domain. Moreover, the approach does not require any intermediate representation of speech; we merely employ the logarithmic mel-scaled spectrograms for evaluation purposes. Achieved correlation for our best participant are comparable to previous results from perceived speech [15] and in the reconstruction of spatio-temporal features of speech [16].

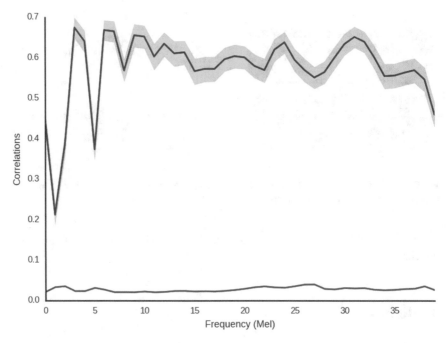

Fig. 3 Reconstruction results for best participant: correlations for each spectral coefficient of our approach (blue) and randomized baseline (red). Reconstruction is consistently better than baseline for all spectral coefficients. Shaded areas denote standard errors

Clearly, our current experiments are carried out in motor-intact participants, for whom the approach provides no additional benefit. However, recent advances in motor decoding from M1 in paralyzed patients [27] give hope that our approach might extend to attempted speech in paralyzed patients, too.

In summary, we present a simple pattern matching approach inspired by the speech synthesis community that is able to reconstruct intelligible speech from intracranial recordings in motor, pre-motor and inferior frontal cortices.

References

1. J.E. Huggins, C. Guger, M. Ziat, T.O. Zander, D. Taylor, M. Tangermann, G. Ruffini, Workshops of the sixth international brain-computer interface meeting: brain–computer interfaces past, present, and future. Brain-Comput. Interf. **4**(1–2), 3–36 (2017)
2. E.M. Mugler, M.C. Tate, K. Livescu, J.W. Templer, M.A. Goldrick, M.W. Slutzky, Differential representation of articulatory gestures and phonemes in precentral and inferior frontal gyri. J. Neurosci. **38**(46), 9803–9813 (2018)
3. J. Chartier, G.K. Anumanchipalli, K. Johnson, E.F. Chang, Encoding of articulatory kinematic trajectories in human speech sensorimotor cortex. Neuron **98**(5), 1042–1054 (2018)

4. F. Lotte, J.S. Brumberg, P. Brunner, A. Gunduz, A.L. Ritaccio, C. Guan, G. Schalk, Electrocorticographic representations of segmental features in continuous speech. Front. Human Neurosci. **9**, 97 (2015)
5. E.M. Mugler, J.L. Patton, R.D. Flint, Z.A. Wright, S.U. Schuele, J. Rosenow, M.W. Slutzky, Direct classification of all American English phonemes using signals from functional speech motor cortex. J. Neural Eng. **11**(3), 035015 (2014)
6. N.F. Ramsey, E. Salari, E.J. Aarnoutse, M.J. Vansteensel, M.G. Bleichner, Z.V. Freudenburg, Decoding spoken phonemes from sensorimotor cortex with high-density ECoG grids. NeuroImage **180**, 301–311 (2018)
7. G. Milsap, M. Collard, C. Coogan, Q. Rabbani, Y. Wang, N.E. Crone Keyword spotting using human electrocorticographic recordings. Front. Human Neurosci. (2019)
8. C. Herff, D. Heger, A. De Pesters, D. Telaar, P. Brunner, G. Schalk, T. Schultz, Brain-to-text: decoding spoken phrases from phone representations in the brain. Front. Neurosci. **9**, 217 (2015)
9. C. Herff, T. Schultz, Automatic speech recognition from neural signals: a focused review. Front. Neurosci. **10**, 429 (2016)
10. D.A. Moses, N. Mesgarani, M.K. Leonard, E.F. Chang, Neural speech recognition: continuous phoneme decoding using spatiotemporal representations of human cortical activity. J. Neural Eng. **13**(5), 056004 (2016)
11. T. Schultz, M. Wand, T. Hueber, D.J. Krusienski, C. Herff, J.S. Brumberg, Biosignal-based spoken communication: a survey. IEEE/ACM Trans. Audio, Speech, Lang. Process. **25**(12), 2257–2271 (2017)
12. S. Chakrabarti, H.M. Sandberg, J.S. Brumberg, D.J. Krusienski, Progress in speech decoding from the electrocorticogram. Biomed. Eng. Lett. **5**(1), 10–21 (2015)
13. A. Stuart, J. Kalinowski, M.P. Rastatter, K. Lynch, Effect of delayed auditory feedback on normal speakers at two speech rates. J. Acoust. Soc. Am. **111**(5), 2237–2241 (2002)
14. F.H. Guenther, J.S. Brumberg, E.J. Wright, A. Nieto-Castanon, J.A. Tourville, M. Panko, P. Ehirim, A wireless brain-machine interface for real-time speech synthesis. PLoS ONE **4**(12), e8218 (2009)
15. B.N. Pasley, S.V. David, N. Mesgarani, A. Flinker, S.A. Shamma, N.E. Crone, E.F. Chang, Reconstructing speech from human auditory cortex. PLoS Biol. **10**(1), e1001251 (2012)
16. S. Martin, P. Brunner, C. Holdgraf, H.J. Heinze, N.E. Crone, J. Rieger, B.N. Pasley, Decoding spectrotemporal features of overt and covert speech from the human cortex. Front. Neuroeng. **7**, 14 (2014)
17. C. Herff, G. Johnson, L. Diener, J. Shih, D. Krusienski, T. Schultz, Towards direct speech synthesis from ECoG: a pilot study, in Engineering in Medicine and Biology Society (EMBC), 2016 IEEE 38th Annual International Conference of the (pp. 1540–1543). IEEE (2016, August)
18. H. Akbari, B. Khalighinejad, J. Herrero, A. Mehta, N. Mesgarani, Towards reconstructing intelligible speech from the human auditory cortex. Scient. Rep. **9**, 874 (2019)
19. M. Angrick, C. Herff, E. Mugler, M.C. Tate, M.W. Slutzky, D.J. Krusienski, T. Schultz, Speech synthesis from ECoG using densely connected 3d convolutional neural networks. J. Neural. Eng. **16**(3), 036019 (2019)
20. M. Angrick, C. Herff, G. Johnson, J. Shih, D. Krusienski, T. Schultz, Interpretation of convolutional neural networks for speech regression from electrocorticography. *ESANN 2018*
21. G.K. Anumanchipalli, J. Chartier, E.F. Chang, Intelligible speech synthesis from neural decoding of spoken sentences. bioRxiv, 481267 (2018)
22. C. Herff, L. Diener, M. Angrick, E. Mugler, M.C. Tate, M.A. Goldrick, D.J. Krusienski, M.W. Slutzky, T.Schultz, Generating natural, Intelligible speech from brain activity in motor, premotor, and inferior frontal cortices. Front. Neurosci. vol. 13, (2019)
23. J.S. Brumberg, D.J. Krusienski, S. Chakrabarti, A. Gunduz, P. Brunner, A.L. Ritaccio, G. Schalk, Spatio-temporal progression of cortical activity related to continuous overt and covert speech production in a reading task. PLoS ONE **11**(11), e0166872 (2016)
24. L. Izhikevich, R. Gao, E. Peterson, B. Voytek, Measuring the average power of neural oscillations. bioRxiv, 441626 (2018)

25. A.J. Hunt, A.W. Black, Unit selection in a concatenative speech synthesis system using a large speech database, in 1996 IEEE International Conference on Acoustics, Speech, and Signal Processing, 1996. ICASSP-96. Conference Proceedings. (Vol. 1, pp. 373–376). IEEE (1996, May)

26. S.S. Stevens, J. Volkmann, E.B. Newman, A scale for the measurement of the psychological magnitude pitch. J. Acoust. Soc. Am. **8**(3), 185–190 (1937)

27. S.D. Stavisky, F.R. Willett, G. H. Wilson, B.A. Murphy, P. Rezaii, D.T. Avansino, W.D. Memberg, J.P. Miller, R.F. Kirsch, L.R. Hochberg, A.B. Ajiboye, S.Druckmann, K.V. Shenoy, J.M. Henderson, Neural ensemble dynamics in dorsal motor cortex during speech in people with paralysis. eLife vol. 8 (2019)

28. L.R. Hochberg, D. Bacher, B. Jarosiewicz, N.Y. Masse, J.D. Simeral, J. Vogel, J.P. Donoghue, Reach and grasp by people with tetraplegia using a neurally controlled robotic arm. Nature **485**(7398), 372 (2012)

Neural Decoding of Attentional Selection in Multi-speaker Environments Without Access to Clean Sources

James O'Sullivan, Zhuo Chen, Jose Herrero, Sameer A. Sheth, Guy McKhann, Ashesh D. Mehta, and Nima Mesgarani

Abstract People who suffer from hearing impairments can find it difficult to follow a conversation in a multi-speaker environment. Modern hearing aids can suppress background noise; however, there is little that can be done to help a user attend to a single conversation without knowing which speaker is being attended to. Cognitively controlled hearing aids that use auditory attention decoding (AAD) methods are the next step in offering help. A number of challenges exist, including the lack of access to the clean sound sources in the environment with which to compare with the neural signals. We propose a novel framework that combines single-channel speech separation algorithms with AAD. We present an end-to-end system that (1) receives a single audio channel containing a mixture of speakers that is heard by a listener along with the listener's neural signals, (2) automatically separates the individual speakers in the mixture, (3) determines the attended speaker, and (4) amplifies the attended speaker's voice to assist the listener. Using invasive electrophysiology recordings, our system is able to decode the attention of a subject and detect switches in attention using only the mixed audio. We also identified the regions of the auditory cortex that contribute to AAD. Our quality assessment of the modified audio demonstrates a significant improvement in both subjective and objective speech quality measures. Our novel framework for AAD bridges the gap between the most recent advancements in speech processing technologies and speech prosthesis research and moves us closer to the development of cognitively controlled hearing aids.

Research supported by NIH, NIDCD, DC014279.

J. O'Sullivan (✉) · Z. Chen · N. Mesgarani
Department of Electrical Engineering, Columbia University, New York, NY, USA
e-mail: jo2472@columbia.edu

S. A. Sheth · G. McKhann
Department of Neurological Surgery, The Neurological Institute, 710 West 168 Street, New York, NY, USA

J. Herrero · A. D. Mehta
Department of Neurosurgery, Hofstra-Northwell School of Medicine and Feinstein Institute for Medical Research, Manhasset, New York, NY, USA

© The Author(s), under exclusive license to Springer Nature Switzerland AG 2020
C. Guger et al. (eds.), *Brain–Computer Interface Research*,
SpringerBriefs in Electrical and Computer Engineering,
https://doi.org/10.1007/978-3-030-49583-1_6

Keywords Auditory attention decoding (AAD) · Speech prosthesis ·
Electrocorticography (ECoG) · Speech separation · Cognitively controlled hearing
aids · Speech quality

1 Introduction

Listening to a single speaker in a multi-speaker environment is extremely challenging
for people who suffer from hearing impairments, which has been attributed to an
increase in listening effort and a reliance on higher-level compensatory cognitive
processes [1]. Assistive hearing devices can suppress certain types of background
noise [2], but they cannot help a user attend to a single conversation without knowing
which speaker is being attending to. Several studies have revealed a dynamic and
selective representation of an attended speaker in human auditory cortex [3–5]. These
findings have led to the idea of auditory attention decoding (AAD): the ability to
decode the identity of an attended speaker over short enough time-scales so as to be
useful for a hearing aid. AAD has been successfully implemented using various neural
signal acquisition methods [3, 6]. However, many challenges must be overcome
before AAD can be practically implemented [7–9], one of which is the lack of
access to the clean sound sources in the environment. One way to address this issue
is beamforming, which uses multichannel microphone recordings to create a neuro-
steered spatial audio filter [7, 10]. However, such an approach is limited to scenarios
where the target and interfering sources are separated in space—a condition that is
not guaranteed.

   We propose a novel framework that combines advances in single-channel speech
separation methods with AAD to alleviate the requirement of a spatial separation
between the target and interfering speakers (Fig. 1). This method can be used in
place of beamforming, or in tandem with it, to create a more realistic solution. Our
method requires prior training on target speakers, meaning that its use is restricted to
a known set of speakers with whom the user interacts. However, new speakers can be
added to this set using a small amount of training data (~20 min). Our system is based
on deep neural network (DNN) audio source separation algorithms [11]. We tested the
efficacy of our system using invasive electrocorticography (ECoG) recordings from
neurological subjects undergoing epilepsy surgery [5]. The high signal-to-noise ratio
of ECoG enabled us to test the upper bound of decoding accuracy and speed, and to
discover which brain regions contributed to the identification of an attended speaker.
This framework for AAD systems bridges the gap between the latest developments in
speech separation algorithms and speech prostheses to help a user attend to a single
speaker in a multi-speaker environment.

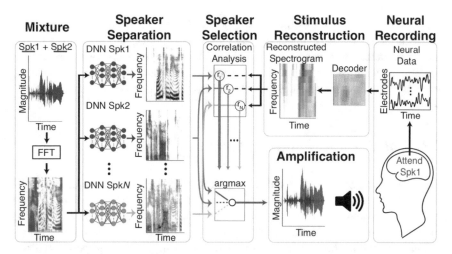

Fig. 1 System schematic. Two speakers, Spk1 (red) and Spk2 (blue), are mixed in a single channel. First, a spectrogram of the mixture is obtained (speakers are marked red and blue for visualization purposes only). The spectrogram is then input to each of several DNNs, each trained to separate a specific speaker. Simultaneously, a user attends to one speaker (Spk1). A spectrogram of this speaker is reconstructed from the user's neural data, which is compared with the outputs of each DNN using a correlation analysis. The appropriate spectrogram is then converted into an acoustic waveform and amplified

2 Methods

2.1 Subjects

6 subjects undergoing clinical treatment for epilepsy took part in this study. All subjects gave their written informed consent to participate in research. 5 subjects were situated at North Shore University Hospital (NSUH), and 1 at Columbia University Medical Center (CUMC). Two subjects (1 and 2) were implanted with high-density electrode arrays over the left temporal lobe. The remaining 4 subjects were implanted with depth electrodes, resulting in varying amounts of coverage over the left and right auditory cortices.

2.2 Stimuli and Experiments

Each subject partook in two experiments: a single-speaker (S-S) and multi-speaker (M-S) experiment. Each subject listened to 4 stories read by a female and male speaker (hereafter referred to as $Spk1_F$ and $Spk2_M$, respectively).

To ensure attentional engagement, the stories were randomly paused and the subjects were instructed to repeat the last sentence. For the M-S experiment, subjects were presented with a mixture of Spk1$_F$ and Spk2$_M$, with no spatial separation between them. The experiment was divided into 4 blocks, and the subjects alternated their attention between the 2 speakers. The stories were randomly paused and the subjects were asked to repeat the last sentence of the attended speaker. All stimuli were presented using a single Bose® SoundLink® Mini 2 speaker.

2.3 Data Preprocessing and Hardware

The subjects at NSUH were recorded using TDT® hardware and the subject at CUMC was recorded using Xltek® hardware. DC drift was removed using a high-pass filter at 1 Hz and data were re-referenced offline using a common average scheme. A period of silence was recorded before both experiments, and all data were normalized (z-scored) relative to this pre-stimulus period. We then obtained the power (Hilbert envelope) of the high gamma (HG) frequency band (70–150 Hz) [5].

2.4 Single-Channel Speaker Separation

To automatically separate each speaker from the mixture, we employed a method of single-channel speech separation that utilizes a class of DNNs known as *long short-term memory* (LSTM) DNNs [11]. Each DNN was trained to separate one specific speaker from arbitrary mixtures. In our experiment, there were only two speakers presented to each subject. However, we are proposing a system that could work in a real-world situation where a device would contain multiple DNNs. Because of this, we trained 4 DNNs to separate 4 speakers, hereafter referred to as Spk1$_F$, Spk2$_M$, Spk3$_F$, and Spk4$_M$. All speakers were native American English speakers. Spk3$_F$ and Spk4$_M$ were taken from the Wall Street Journal (WSJ) corpus.

The speech waveforms were converted into 100-dimensional Mel-frequency spectrograms. The goal was then to obtain an estimate $\hat{S}$ of a clean target spectrogram S from a mixture M. To do so, a soft mask $\hat{Y}$ was learnt and applied to the mixture to mask the interfering speech. The squared Euclidian distance between the masked spectrogram and the clean target spectrogram was treated as the error in order to generate the gradient that was back propagated through the DNN to update the parameters. Each DNN had 4 layers with 300 nodes each, followed by a single layer with 100 nodes in order to output a spectrogram. An acoustic waveform was generated by combining this spectrogram with the phase of the original mixture. See [11] for further information. For training, we used twenty minutes of speech from the target speakers and ~5 h of speech from 103 interfering speakers from the WSJ corpus. The target speaker was always mixed with one interfering speaker with the same RMS intensity. Unseen utterances were used for testing (for both the target and interfering speakers). The DNNs never saw any of the other target speakers during training.

2.5   Speaker-Separation Performance

Speaker-separation performance was measured in 3 ways: (1) by obtaining the correlation between the DNN output and the spectrogram of the clean target speaker, and (2) by using an objective measure of speech quality known as the Perceptual Evaluation of Speech Quality (PESQ) score. The PESQ algorithm produces a score between 1.0 and 4.5, where higher values indicate better quality. (3) We also asked 12 naïve participants to rate (from 1 to 5; mean opinion score; MOS) listening effort in attending to a speaker in the mixture when that speaker was amplified (system on) or not (system off).

2.6   Stimulus-Reconstruction

To determine the attended speaker, we employed a method known as stimulus-reconstruction [5], which applies a spatiotemporal filter (decoder) to neural recordings to reconstruct a spectrogram of an attended speaker. See [5] for further information.

2.7   Neural Correlation Analysis

Determining to whom the subject is attending requires a correlation analysis. Typically, whichever spectrogram has the largest correlation with the reconstructed spectrogram is taken to be the attended speaker [6, 8]. However, because we are using 4 DNNs, each trained to separate a different speaker, the analysis becomes more complex. Crucially, it was necessary to normalize the correlation values with respect to the mixture, because the correlation between the reconstructed spectrograms and the mixture was very large (see Results; Fig. 3).

   For clarity we will first define some terminology: a spectrogram outputted from the kth DNN will be referred to as S_{DNN_k}, the spectrogram of the mixture as S_{MIX}, and the reconstructed spectrogram as S_{RECON}. In order to emphasize large correlations, we applied a Fisher transformation (inverse hyperbolic tangent) to each r-value. The normalization procedure involved five steps. First, we obtained the correlation between S_{RECON} and each S_{DNN_k}, which we will refer to as ρ_{1_k}. Next, we obtained the correlation between S_{RECON} and the difference between S_{DNN_k} and S_{MIX}, which we will refer to as ρ_{2_k}. Intuitively, this value should be close to zero if a DNN is outputting the mixture, small if a DNN is correctly separating the attended speaker, and large if it separates the unattended speaker. Therefore, taking the difference of ρ_{1_k} and ρ_{2_k}, and dividing by their sum, should produce a score (α_k) that can differentiate between each case. This was followed by a test-normalization (t-norm), where each α score was normalized relative to the distribution of α scores from all DNNs:

$$\beta_k = \frac{\alpha_k - \mu_\alpha}{\sigma_\alpha}$$

where μ_α and σ_α are the mean and SD of the distribution of α scores. To further penalize DNNs outputting the mixture, we subtracted the correlation between S_{DNN_k} and S_{MIX}, and added the constant 1:

$$Pk = \beta_k - r\left(S_{DNN_k}, S_{MIX}\right) + 1$$

2.8   Attention Decoding Index (ADI)

To obtain a measure of our ability to determine the attended speaker from neural recordings, we segmented the reconstructed spectrograms into 20-s bins and obtained 4 normalized correlation values for each segment: $P1_f$, $P2_m$, $P3_f$, and $P4_m$. To take into account any potential bias for a particular speaker, we define the Attention Decoding Index (ADI) as the proportion of the number of correct hits minus the number of false positives, bounded between $[-1, 1]$. Significant performance was determined to be 0.45 (3 times the standard deviation of a null distribution of ADI obtained via a randomized shuffle of the data.)

2.9   Dynamic Switching of Attention

To simulate a dynamic scenario in which the subjects were switching attention, we divided and concatenated the data into 10 consecutive 60 s segments in which the subjects were attending to either speaker. To track the attentional focus of each subject, we used a sliding window to obtain normalized correlation values each second.

3   Results

3.1   Speaker-Separation Performance

To examine the ability of the DNNs to separate their designated speakers from the mixtures, we measured the correlation between the output of each DNN and the clean target speaker spectrograms (gray bars; Fig. 2a). We also tested performance when the designated speaker was not present in the mixture (red/blue bars). As expected, the networks could not separate undesignated speakers from mixtures. In addition,

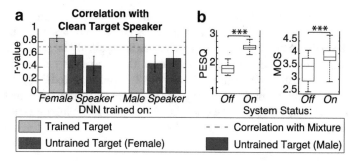

Fig. 2 DNN speaker-separation performance. **a** Results are divided into cases when the DNNs were trained to separate a female (left) or a male (right) speaker. Error bars represent SD. Gray bars show when a DNN was presented with a mixture containing its pre-trained speaker, and the red and blue bars when the mixture contained an undesignated speaker that was female (red) or male (blue). The dotted line shows the average correlation between the raw mixture and the clean target speaker. **b** Objective (PESQ) and subjective (MOS) scores for the raw mixtures (system off) and the outputs of the DNNs (system on)

the output of the system produced objectively and subjectively cleaner speech signals, with significant increases in both the PESQ and MOS scores (Wilcoxon signed-rank test, $p < 0.001$; Fig. 2b).

3.2 Neural Correlation Analysis

To determine which speaker a subject was attending to, we performed a neural correlation analysis where we compared the reconstructed spectrograms (from the neural data) with the output of each DNN (Fig. 3). The left panel (raw) shows the average correlation between the reconstructed spectrograms and the outputs of the DNNs for each subject. Because the subjects alternated their attention between two speakers, the

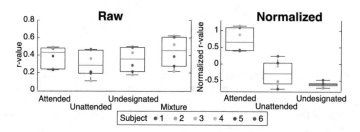

Fig. 3 Reconstruction accuracy. The correlations between the reconstructed spectrograms and the outputs of the DNNs. The left (right) panel shows the raw (normalized) r-values. Each subject is represented by a colored dot. Because the subjects alternated their attention between two speakers, the r-values labeled as attended and unattended come from the DNNs trained on $Spk1_F$ and $Spk2_M$, and the r-values labeled as undesignated come from the DNNs trained on $Spk3_F$ and $Spk4_M$

r-values labeled as *attended* and *unattended* come from the DNNs trained on $Spk1_F$ and $Spk2_M$. The r-values labeled as *undesignated* come from the DNNs trained on $Spk3_F$ and $Spk4_M$. Although the attended r-values are typically larger than the unattended r-values, there is also a large correlation with the mixture, and therefore with the DNNs that were not trained on $Spk1_F$ and $Spk2_M$ (undesignated). This is because these DNNs typically outputted spectrograms that were very close to the mixture. To account for this, we normalized the r-values with respect to the mixture (right panel; see methods).

3.3   Attention Decoding Index

It was possible to decode the attentional focus of 3 subjects (1, 2, and 3; Fig. 4a). There was no significant difference in ADI when using the ideal spectrograms (Wilcoxon signed-rank test; $p > 0.05$). We sought to explain the variability in ADI across subjects by identifying the anatomical locations of each electrode: the pie charts above each subject (Fig. 4a) illustrate the proportion of electrodes from 2 anatomical regions (see figure caption). Electrodes in STG and Other produced ADIs significantly greater than zero (Fig. 2b; $p < 0.001$), compared to the use of electrodes in HG ($p = 0.02$).

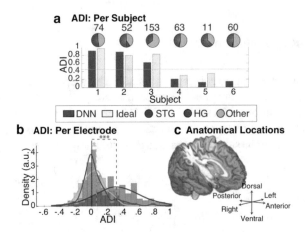

Fig. 4 Attention decoding index (ADI). **a** The proportion of segments (20 s) in which the attentional focus of each subject could be correctly determined. The gray line indicates an ADI significantly above chance (0.45; see methods). The pie charts illustrate the proportion of electrodes from 2 anatomical regions: Heschl's Gyrus (HG; red) and Superior Temporal Gyrus (STG; blue). Electrodes responsive to speech, but not in either of these locations, are referred to as Other (green). The single number above each pie chart refers to the total number of electrodes that were responsive to speech for that subject. **b** We also obtained the ADI for each individual electrode. Bars are colored according to anatomical location. **c** Visualization of the anatomical locations HG (red), STG (blue) and Other (green), from an example subject (subject 4). All brain regions above the lateral sulcus in the right hemisphere have been removed to expose HG

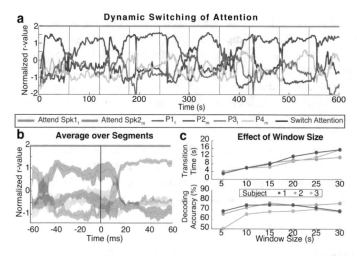

Fig. 5 **a** Dynamic switching of attention for subject 1. Black lines indicate a switch in attention, and the colored bar on top indicates the speaker being attended to. Beneath are the normalized correlation values for each of the 4 DNNs plotted over time. **b** The same data as in *A*, but with the average of all segments in which the male (female) speaker was attended on the left (right) of the black line. Shaded regions denote standard error. (C) The data in *A* and *B* were obtained using a 20 s window. Here we display the decoding accuracies and transition times obtained using a range of window-sizes for each subject whose attention could be decoded

3.4 Dynamic Switching of Attention

Figure 5a shows the results when simulating a dynamic switching of attention for an example subject (subject 1) using a 20 s window size. Figure 5b displays the same results but averaged over all sections when the subject was attending to $Spk2_m$ (-60 s:0 s) and $Spk1_f$ (0 s:60 s).

Figure 5c shows how changing the window size affects decoding-accuracy and transition time (how long it takes to detect a switch in attention) for each subject whose attention we could decode. The transition times were calculated as the time at which the blue ($Spk2_M$) and red ($Spk1_F$) lines intersect in the averaged data (e.g., Figure 3b).

4 Discussion

We have developed an end-to-end system that incorporates the latest single-channel automatic speech-separation algorithms into the auditory attention-decoding (AAD) platform. In addition to successfully identifying the attended speaker, our system also amplifies that speaker, resulting in a significant increase in the subjective quality of the listening experience. Demos of the final audio output as a subject switches attention are provided online [12]. We also determined that STG is important for successfully decoding attention. This is an important finding for non-invasive AAD research where source localization methods can be used to target specific brain regions.

A practical limitation for all algorithms intended for hearing aids is that hardware constraints could limit the number of DNNs that could be housed inside a portable device. However, modern hearing aids are able to perform off-board computing by interfacing with a cell phone [2]. Another consideration is the fact that DNNs rely heavily on the data used to train them. Therefore, additional training would be required to separate speakers under different environmental conditions [13]. Also, because people tend to involuntarily speak louder in noisy situations, which affects acoustic features such as pitch, rate and syllable duration (the Lombard effect), this would also need to be taken into account during training of the DNNs.

References

1. J.E. Peelle, A. Wingfield, The neural consequences of age-related hearing loss, Trends Neurosci. (2016)
2. J.L. Clark, D.W. Swanepoel, Technology for hearing loss–as we know it, and as we dream it. Disab. Rehabil. Assist. Tech. **9**, 408–413 (2014)
3. N. Ding, J.Z. Simon, Emergence of neural encoding of auditory objects while listening to competing speakers. Proc. Natl. Acad. Sci. U.S.A. **109**, 11854–11859 (2012)
4. A.J. Power, J.J. Foxe, E.J. Forde, R.B. Reilly, E.C. Lalor, At what time is the cocktail party? A late locus of selective attention to natural speech. Eur. J. Neurosci. **35**, 1497–1503 (2012)
5. N. Mesgarani, E.F. Chang, Selective cortical representation of attended speaker in multi-talker speech perception. Nature **485**, 233-U118 (2012)
6. J.A. O'Sullivan, A.J. Power, N. Mesgarani, S. Rajaram, J.J. Foxe, B.G. Shinn-Cunningham et al., Attentional selection in a cocktail party environment can be decoded from single-trial EEG. Cerebral Cortex **25**, 1697–1706 (2015)
7. S. Van Eyndhoven, T. Francart, A. Bertrand, EEG-informed attended speaker extraction from recorded speech mixtures with application in neuro-steered hearing prostheses. *arXiv preprint* arXiv:1602.05702 (2016)
8. B. Mirkovic, S. Debener, M. Jaeger, M. De Vos, Decoding the attended speech stream with multi-channel EEG: implications for online, daily-life applications. J. Neural Eng. **12**, 046007 (2015)
9. M.G. Bleichner, B. Mirkovic, S. Debener, Identifying auditory attention with ear-EEG: cEEGrid versus high-density cap-EEG comparison. J. Neural Eng. **13**, 066004 (2016)
10. N. Das, S. Van Eyndhoven, T. Francart, A. Bertrand, Adaptive attention-driven speech enhancement for EEG-informed hearing prostheses, in *2016 IEEE 38th Annual International Conference of the Engineering in Medicine and Biology Society (EMBC)* (2016) pp. 77–80
11. F. Weninger, J.R. Hershey, J. Le Roux, B. Schuller, Discriminatively trained recurrent neural networks for single-channel speech separation, in *IEEE Global Conference on Signal and Information Processing (GlobalSIP)*, pp. 577–581 (2014)
12. http://naplab.ee.columbia.edu/nnaad.html
13. J. Li, L. Deng, Y. Gong, R. Haeb-Umbach, An overview of noise-robust automatic speech recognition. IEEE/ACM Trans. Audio, Speech, Lang. Process. **22**, 745–777 (2014)

Goal-Directed BCI Feedback Using Cortical Microstimulation

Yohannes Ghenbot, Xilin Liu, Han Hao, Cole Rinehart, Sam DeLuccia, Solymar Torres Maldonado, Gregory Boyek, Milin Zhang, Firooz Aflatouni, Jan Van der Spiegel, Timothy H. Lucas, and Andrew G. Richardson

Abstract Paralyzed individuals would benefit from brain-computer interface (BCI) systems that restore not just motor function but also tactile and proprioceptive feedback. Such feedback has been shown to be critical to motor performance. Intracortical microstimulation (ICMS) has often been employed to provide artificial sensory feedback. However, it remains a question of how best to encode the multidimensional nature of this information (e.g. location, intensity, frequency of tactile signals). This project explored encoding goal-directed error signals as a way to simplify the feedback. We used a behavioral paradigm with rats in which ICMS was used as a tunable error signal to direct the subjects to unseen goal locations. We found that with relatively little training, the rats performance in the task with ICMS feedback was statistically as good as with natural sensory feedback. The results provide a demonstration that multidimensional sensory feedback can be mapped to single goal-related encoded signal in certain behavioral contexts to decrease the cognitive burden associated with interpreting multiple ICMS-evoked percepts.

Keywords Brain-computer interface (BCI) · Intracortical microstimulation (ICMS) · Sensory feedback

Y. Ghenbot · C. Rinehart · S. DeLuccia · S. T. Maldonado · G. Boyek · T. H. Lucas · A. G. Richardson (✉)
Department of Neurosurgery and Center for Neuroengineering and Therapeutics, University of Pennsylvania, Philadelphia, PA, USA
e-mail: andrew.richardson@pennmedicine.upenn.edu

X. Liu · H. Hao · M. Zhang · F. Aflatouni · J. Van der Spiegel
Department of Electrical and Systems Engineering, University of Pennsylvania, Philadelphia, PA, USA

© The Author(s), under exclusive license to Springer Nature Switzerland AG 2020
C. Guger et al. (eds.), *Brain–Computer Interface Research*,
SpringerBriefs in Electrical and Computer Engineering,
https://doi.org/10.1007/978-3-030-49583-1_7

1   Introduction

1.1   Paralysis and Motor BCI

An estimated 5.4 million United States citizens (approximately 2%) live with some degree of paralysis as a result of CNS insult—primarily stroke and spinal cord injury [1]. Several emerging therapeutics are under investigation to restore mobility to paralyzed patients. Motor brain computer interface (BCI) technology has been extensively investigated as a strategy to replace lost movement abilities [2]. Motor BCIs bypass damaged neural tracks, allowing action intention signals recorded from intact cortical motor areas to command external actuators (e.g. cursor or myoelectric prosthesis) or the paralyzed limb directly through functional electrical stimulation [3].

Motor BCI technology is currently limited by a lack of tactile and proprioceptive sensory feedback, which is also disrupted in paralysis. Indeed, even in sensory deafferented states where volitional movement is preserved, skilled motor performance has been shown to degrade and not improve with time and training [4]. Thus, researchers are quantifying the sensory percepts elicited by intracortical microstimulation (ICMS) and developing strategies to incorporate this artificial feedback into closed-loop BCI paradigms [5].

1.2   Closed-Loop Sensory Feedback Strategies

Somatosensory information is complex and multidimensional, originating from distributed skin, muscle, and joint mechanoreceptors with a wide range of sensitivities. Artificial replication of this information for BCI applications is a challenging problem. Most studies have taken a straightforward biomimetic approach, replacing a missing sense (e.g. fingertip force) with microstimulation of the brain area normally encoding that sense (e.g. primary somatosensory cortex, S1) [6]. However, given the complexity of the feedback and potential cognitive burden of interpreting multiple artificially-derived sensory percepts, the biomimetic strategy is not necessarily scalable. Investigations into alternative encoding strategies are warranted.

As an alternative approach, we posit: (a) some BCI-controlled actions or subactions have a known goal that is dependent on some measurable aspect of the environment and (b) that a scalar function of the actions and measurements can be derived whose value represents the deviation from the goal. In these cases, goal-directed BCI actions can be guided simply by a one-dimensional map from the deviation value to cortical stimulation. Although the feedback may not correspond with any natural sense, we hypothesize that it would be intuitive for a user to adjust actions to minimize the deviation and reach the goal. The closed-loop BCI system thus would operate akin to a simple servo-controlled mechanism (e.g. thermostat). Our rationale for this approach is that it places the burden of interpreting multiple sensory signals on the BCI hardware rather than on the brain, in contrast to the biomimetic approach.

To test this hypothesis, we designed a novel searching task for rats [7]. Instead of conveying overly detailed information about goal locations, we informed the rats of their heading relative to the straight path to the goal using ICMS. While in our prior study we focused on task learning with this error-related feedback, here we reanalyzed the data to quantify the plateau performance (i.e. shortest path to goal) achieved in natural and artificial sensory feedback conditions.

2 Methods

2.1 Paradigm

Our experiment used adult male Sprague-Dawley rats ($N = 6$) and was approved by the Institutional Animal Care and Use Committee of the University of Pennsylvania. Five rats (Sa, Sn, Ro, Ge, Fr) were implanted unilaterally with a concentric bipolar stimulating electrode in S1. One rat (Mk) was implanted in A1 to investigate encoding outside of S1. During testing, the electrode was connected to a custom wireless neural stimulator [8]. On each trial, the rat was placed at the center of a 2-m diameter pool and had to swim to an invisible, submerged platform (Fig. 1a). Unlike the classic Morris water maze task [9], in our experiment, the platform was moved to a random location on each trial so that a visually-cued memory of platform location could

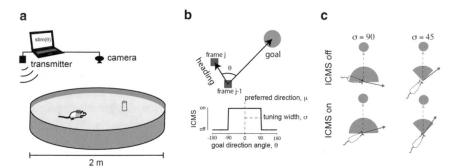

Fig. 1 Experimental paradigm. **a** Rats implanted with a stimulating electrode in sensory cortex and wearing a wireless neural stimulator (red box) were placed in a water maze with a hidden, submerged platform. An overhead camera tracked the swim path and updated stimulation parameters in real time. **b** ICMS was delivered as a function of the rat's instantaneous goal direction angle (θ). The step encoding function was defined by a preferred direction (μ) and tuning width (σ). **c** Illustration of conditions in which ICMS was on or off for two different tuning width values. A rat received ICMS when its heading (red arrow) was within a range (green sector) whose width was defined by σ and whose orientation relative to the platform (green circle) was defined by μ. Figure adapted with permission from [7]

Table 1 ICMS encoding function parameters for each rat

σ (deg)	μ (deg)	rat
5	0	Ge, Fr
15	0	Ge, Fr
45	0	Ge, Fr, Mk
90	0	Ge, Sa
90	180	Sn, Ro, Sa

not be formed. Indeed, prior work has shown that rats are not able to improve their performance on this random-location task using their natural senses [10]. First, the rats were trained on the task using a visual cue attached to the platform that rose above the surface of the water for baseline performance. In subsequent trials, the rats were given platform directional information through ICMS of sensory cortex. Catch trials were intermittently incorporated, in which the ICMS was turned off. For two rats, obstacles were added to the behavioral arena after achieving plateau performance to increase task complexity and to investigate the dependency on continuous sensory feedback.

2.2 ICMS Parameters

The rat's swim path was monitored by a video camera at 12 frames/s. On each frame, custom software computed the goal direction angle, θ, on the basis of the rat's current heading relative to the direction of the platform and wirelessly updated ICMS parameters as a function of this angle (Fig. 1b). The goal encoding function was defined by a preferred direction (μ) and tuning width (σ). When $\theta - \mu \leq \sigma$, we delivered 100-Hz trains of charge-balanced, 0.2-μs duration current pulses at suprathreshold intensity (15–75 μA, measured by evoked behavioral response prior to water maze experiments). When $\theta - \mu > \sigma$, we provided no stimulation. The binary, rather than continuous, function output ensured that the stimulus was felt by the rat throughout the entire tuning width. The tuning width parameter provided control over the acuity of the encoded goal direction information (Fig. 1c). The sensory encoding conditions tested are listed in Table 1.

3 Results

Each rat performed a series of trial blocks (127 ± 49 trials per block on average), where each block was defined by a specific sensory feedback condition. Performance on each trial was quantified by the path length ratio, which was the ratio of the swim path length to the straight-line path between the start and platform locations. Within each block, the rats typically exhibited learning (i.e. decreasing path length ratio

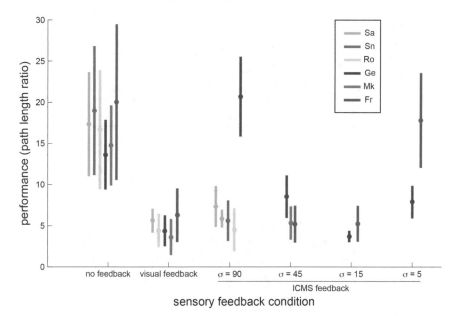

Fig. 2 Plateau performance of rats under each tested sensory condition: no feedback, visual feedback, and ICMS feedback. Shown are the mean and 95% confidence interval of the path length ratio after learning

values) until they reached a plateau level of performance [7]. Plateau performance was calculated as the average path length ratio in the final 10–30 trials of each block. Results are summarized in Fig. 2.

Rat Ge was tested over a wide range of encoding parameters (Table 1). Ge was unable to learn the task using the low acuity encoding parameter of $\sigma = 90°$ after 85 trials, as performance was actually worse than that of catch (i.e. no feedback) trials (Table 2). When directional feedback was less ambiguous ($\sigma \leq 45°$), Ge was able to utilize the goal-direction signals for significantly enhanced task performance (Fig. 2). Performance was best at a tuning width of $\sigma = 15°$. Further decreases in acuity of ICMS feedback ($\sigma = 5°$) resulted in worse performance (Fig. 2).

Rat Fr learned to perform the task with ICMS feedback ($\sigma = 45°$ and 15°) at a proficiency equal to that with visual feedback. However, similar to rat Ge, Fr's performance worsened under the 5° tuning width. Thus, there was an optimal encoding function for conveying goal direction information with ICMS. Furthermore, performance was dependent on the acuity of the goal-direction information, but not upon the site of stimulation. The encoded information remained useful in other primary sensory areas, as rat Mk learned to use the artificial directional information when encoded into A1.

In order to further test the rats' proficiency in using ICMS feedback, we incorporated obstacles into the water maze. The platform was located behind the obstacles during only half of these trials to ensure that rats did not use the barrier as a visual cue

Table 2 Statistical analysis of behavioral performance

σ (deg)	μ (deg)	Rat	ICMS versus no feedback	ICMS versus visual feedback
5	0	Ge	$t(48) = -3.18, p = 0.0026$	
		Fr	$t(44) = -0.438, p = 0.664$	
15	0	Ge	$t(47) = -5.95, p = 3.26 \times 10^{-7}$	$t(58) = 0.914, p = 0.365$
		Fr	$t(44) = -4.16, p = 1.44 \times 10^{-4}$	$t(43) = -0.0788, p = 0.938$
45	0	Ge	$t(47) = -1.41, p = 0.165$	
		Fr	$t(44) = -4.48, p = 5.22 \times 10^{-5}$	
		Mk	$t(34) = -3.89, p = 4.39 \times 10^{-4}$	$t(38) = -0.0833, 0.934$
90	0	Ge	$t(48) = 2.25, p = 0.0292$	
		Sa	$t(45) = -2.97, p = 0.0048$	
90	180	Sn	$t(27) = -2.53, p = 0.0175$	
		Ro	$t(23) = -2.59, p = 0.0162$	$t(16) = -1.78, p = 0.0942$
		Sa	$t(50) = -3.97, p = 2.33 \times 10^{-4}$	$t(39) = -0.368, p = 0.715$
			$t(45) = -2.97, p = 0.0048$	

of platform location. Rats Ge and Fr were still able to complete the task efficiently despite absence of continuous ICMS feedback guiding them to the platform (Fig. 3). This result highlights the skill that rats attained with ample training.

Next, we explored whether rats could utilize low acuity sensory signals that were ineffective in rat Ge (σ = 90°). This is relevant to BCI technology as sensors may be limited in the quality of information that could be delivered to a BCI user. In particular, there may be delays in task relevant event detection and conversion to the appropriate stimulus as well as sensory noise. In the remaining rats (Sn, Ro, and Sa) we explored σ = 90°, with μ = 0° or 180°. At plateau performance, all three rats demonstrated superior performance during ICMS trials when compared to catch trials. We further challenged rat Sa by reversing the preferred direction μ. He was able to learn the new remapping by the end of the trial block (Table 2).

Overall, five rats reached a plateau performance with ICMS feedback that was not statistically different than with visual feedback (Table 2). The impressive performance despite changes in acuity of information was achieved by adopting behavioral strategies that were customized to the high and low acuity sensory conditions. Rats using low acuity feedback adopted a looping strategy to find the platform, while rats that used higher acuity signals performed zigzagging behaviors (Fig. 3).

4   Discussion

We investigated goal encoding as a unique approach to avoid biomimetic scalability issues. Our results show that rats were able to use continuous, non-native, egocentric information concerning the direction of a hidden goal at a proficiency identical to

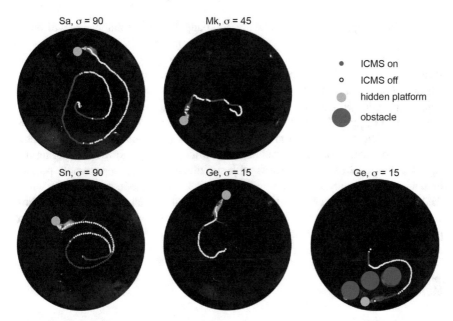

Fig. 3 Example trials showing search strategies tailored to different sensory encoding functions. Plots show final frame of each trial with superimposed graphics indicating swim path, platform, and obstacle locations. With low acuity goal direction encoding ($\sigma = 90°$), rats adopted looping strategies (left column). With high acuity encoding ($\sigma \leq 45°$), rats adopted zigzagging strategies (middle column). Rats were able to circumvent obstacles to find the platform using only minimal initial ICMS feedback (right)

natural vision. Interestingly, despite being placed under different sensory conditions, rats attained similar plateau performances by adopting search strategies tailored to sensory constraints [7].

Similar goal-directed error signals have been demonstrated in a primate reaching task [11] and a human hand aperture task [12]. In the human task, percepts were generated by stimulation of somatosensory cortex using electrocorticographic electrodes. The subject wore a glove that sensed hand aperture, equipped with the ability to measure whether hand aperture was wide, appropriate, or tight. Deviation from the goal aperture was communicated to the user via three stimulus functions—two perceptually discernable stimuli for wide and tight apertures and no stimulation for the appropriate aperture. This information allowed the subject to maintain the appropriate aperture.

In addition to hand aperture, we suggest that goal-encoding feedback could be useful in another critical aspect of grasping: grip force. Grasping tasks require appropriate grip force to lift and transport an object without the object slipping (too little grip force) or getting crushed (too much grip force). Performance on this task degrades in the absence of somatosensation [13]. Sensors on the hand could detect all the appropriate variables: shear forces at each contact point due to object mass and acceleration and change in joint angles after contact due to object compliance. In the biomimetic approach, each force and joint angle could be mapped to a different stimulation site or parameter to approximate the natural condition. However, the parallel stimulus channels would place a high cognitive load on the user. Instead, stimulation at a single site could be driven by a scalar function of these variables, whose value indicates the deviation from the appropriate grip force given the sensed object properties. Binary slip on-off demonstrations with artificial sensors have been proposed to improve dexterity in the field of robotics [14]. Of course, one could use the deviation value to automatically adjust the BCI motor commands without the user's awareness (i.e. implementing an artificial reflex). However, bringing this signal to the level of consciousness with stimulus-evoked percepts would provide both a sense of agency and behavioral flexibility, for example in manual tasks where dexterity improves by effectively decreasing the safety margin against slip. Thus, goal encoding appears to be a viable BCI feedback strategy in a number of different important behavioral contexts.

References

1. B.S. Armour, E.A. Courtney-Long, M.H. Fox, H. Fredine, A. Cahill, Prevalence and causes of paralysis-United States, 2013. Am. J. Public Health **106**(10), 1855–1857 (2016). https://doi. org/10.2105/AJPH.2016.303270
2. M.A. Lebedev, M.A. Nicolelis, Brain-machine interfaces: From basic science to neuropros- theses and neurorehabilitation. Physiol. Rev. **97**(2), 767–837 (2017). https://doi.org/10.1152/ physrev.00027.2016
3. A.B. Ajiboye et al., Restoration of reaching and grasping movements through brain-controlled muscle stimulation in a person with tetraplegia: A proof-of-concept demonstration. Lancet, Mar 28 2017. https://doi.org/10.1016/s0140-6736(17)30601-3
4. A.G. Richardson et al., The effects of acute cortical somatosensory deafferentation on grip force control. Cortex **74**, 1–8 (2016). https://doi.org/10.1016/j.cortex.2015.10.007
5. D.J. Weber, R. Friesen, L.E. Miller, Interfacing the somatosensory system to restore touch and proprioception: Essential considerations (in eng). J. Mot. Behav. **44**(6), 403–418 (2012). https://doi.org/10.1080/00222895.2012.735283
6. G.A. Tabot et al., Restoring the sense of touch with a prosthetic hand through a brain interface (in eng). Proc. Natl. Acad. Sci. U. S. A. **110**(45), 18279–18284 (2013). https://doi.org/10.1073/ pnas.1221113110
7. A.G. Richardson et al., Learning active sensing strategies using a sensory brain-machine inter- face. Proc. Natl. Acad. Sci. U. S. A. **116**(35), 17509–17514 (2019). https://doi.org/10.1073/ pnas.1909953116
8. X. Liu et al., A wireless neuroprosthetic for augmenting perception through modulated electrical stimulation of somatosensory cortex, in *presented at the 2017 IEEE International Symposium on Circuits and Systems (ISCAS)*, 28–31 May 2017, 2017
9. C.V. Vorhees, M.T. Williams, Morris water maze: Procedures for assessing spatial and related forms of learning and memory. Nat. Protoc. **1**(2), 848–858 (2006). https://doi.org/10.1038/ nprot.2006.116
10. E. Baldi, C.A. Lorenzini, B. Corrado, Task solving by procedural strategies in the Morris water maze. Physiol. Behav. **78**(4–5), 785–793 (2003)
11. M.C. Dadarlat, J.E. O'Doherty, P.N. Sabes, A learning-based approach to artificial sensory feedback leads to optimal integration. Nat. Neurosci. **18**(1), 138–144 (2015). https://doi.org/ 10.1038/nn.3883

12. J.A. Cronin et al., Task-specific somatosensory feedback via cortical stimulation in humans. IEEE Trans. Haptics **9**(4), 515–522 (2016). https://doi.org/10.1109/toh.2016.2591952
13. J. Monzee, Y. Lamarre, A.M. Smith, The effects of digital anesthesia on force control using a precision grip. J. Neurophysiol. **89**(2), 672–683 (2003). https://doi.org/10.1152/jn.00434.2001
14. R.A. Romeo, C.M. Oddo, M.C. Carrozza, E. Guglielmelli, L. Zollo, Slippage detection with piezoresistive tactile sensors. Sensors (Basel) **17**(8) (2017). https://doi.org/10.3390/s17081844

Neuromotor Recovery Based on BCI, FES, Virtual Reality and Augmented Feedback for Upper Limbs

Robert Gabriel Lupu, Florina Ungureanu, Oana Ferche, and Alin Moldoveanu

Abstract Recently investigated rehabilitative practices involving Brain-Computer Interface (BCI) and Functional Electrical Stimulation (FES) techniques provided long-lasting benefits after short-term recovering programs. The prevalence of this revolutionary approach received a boost from virtual reality and augmented reality, which contribute to the brain neuroplasticity improvement and can be used in neurorehabilitation and treatment of motor/mental disorders. This work presents a therapy system for stroke rehabilitation based on these techniques. The novelty of the proposed system consists of including an eye tracking device that detects the patient's vigilance during exercises and warns if patient is not focused on the items of interest from the virtual environment. This additional feature improves the level of user involvement and makes him/her conscious of the rehabilitation importance and pace. Moreover, the system architecture is reconfigurable, and the functionalities are specified by software. The laboratory tests have validated the system from a technical point of view, and preliminary results from the clinical tests have highlighted the system's quick accommodation to the proposed therapy and fast progress for each user.

1 Introduction

Rehabilitation is an important part of recovery and helps the patient to become more independent after a stroke or a motor/mental disorder. In the last decade, the Brain-Computer Interface (BCI), the Virtual Reality (VR) and the Functional Electrical Stimulation (FES) techniques are widely used in more complex and efficiently

R. G. Lupu (✉) · F. Ungureanu
Faculty of Automatic Control and Computer Engineering, Computer Engineering Department,
"Gheorghe Asachi" Technical University of Iasi, Dimitrie Mangeron 27, 700050 Iasi, Romania
e-mail: lupu.robert@gmail.com

O. Ferche · A. Moldoveanu
Computer Engineering Department, "Politehnica" University of Bucharest, Bucharest, Romania

© The Author(s), under exclusive license to Springer Nature Switzerland AG 2020
C. Guger et al. (eds.), *Brain–Computer Interface Research*,
SpringerBriefs in Electrical and Computer Engineering,
https://doi.org/10.1007/978-3-030-49583-1_8

systems aiming to bolster the rehabilitation process. In this context, different specific devices became affordable, and many research groups and health institutions are focused on motor, cognitive or speech recovery after stroke (Stroke Centre from Johns Hopkins Institute0, ENIGMA-Stroke Recovery, StrokeBack) [1–3].

In this paper, we present an affordable system for recovery of patients with neuro-motor impairments following strokes, traumas or brain surgery. It relies on a brand new idea—recovery through augmented and magnified feedback—that creates new, distinct possibilities to overcome block stages typical to early recovery, to stimulate recovery through neuroplasticity. The system was customised and tested for upper-limb recovery but can be tailored for any other particular purpose. Another own idea of our approach is that the tasks and guidance are provided by a virtual therapist—a new concept in the field of rehabilitation and considered extremely promising by the healthcare professionals. Besides others research projects dedicated to upper limb recovery (RETRAINER, NIHR) [4, 5] or very recent published works [6, 7], our solution makes use of an eye-tracking method to provide a warning if the patient stops concentrating during exercises.

The purpose of the proposed recovery system is to help in fulfilling the causal chain/loop of recovery, consisting mainly of three steps: motor act is performed or attempted, by the patient, with or without external help; the patient observes sensations and results (visually, haptic or proprioceptive); the patient's cortex associates the motor act with the observations and gradually learns and perfects the motor act. Most techniques and systems for neuromotor recovery only pay attention to the motor act performance, neglecting the essentiality of observation. The system handles the whole recovery causal chain in a unified way. Previous versions and facilities of presented recovery system were designed and implemented in the framework of TRAVEE project [8] and are presented in a comprehensive manner in some published papers [9–12].

From a user's point of view, the system has two main components: one that is dedicated to the patient that undergoes the rehabilitation process after stroke, and one that is dedicated to the therapist—the clinician that guides the rehabilitation session [8]. The complex system dedicated to stroke rehabilitation involves devices and software that immerse the patient in a Virtual Environment to identify themselves with the presented avatar, as well as devices dedicated to support his/her movements and providing complex feedback during the exercises. The component for the therapist is aimed mostly at providing intuitive tools for configuring the rehabilitation session composition and the devices used for each exercise, as well as to monitor the activity of the patient.

2   Materials and Methods

The system is designed to support three main features: patient monitoring, patient training and stimulation and data analysis and processing, Fig. 1a. Devices for the first two features are each optional "plugin" components of the system. Hence, the

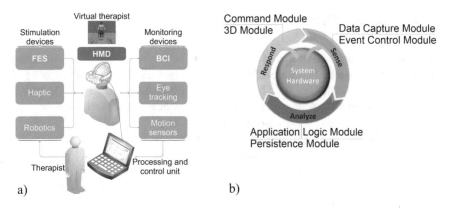

Fig. 1 **a** The system architecture; **b** Sense—analyse—respond mapping functions

system can be tailored to use all of the devices together, but not all of them are mandatory. Results of processed raw data from monitoring devices are used to trigger the stimulation devices with respect to rehabilitation exercise [11]. The software has an event driven architecture to manage needs like real-time data processing, communication and security, data access/storage, patient condition and working conditions, interoperability [9]. The running processes are managed through the sense-analyse-respond approach shown in Fig. 1b. In the "sense" component, monitoring devices capture and process data in real-time. "Analyse" refers to continuous evaluation of the processing results (when clause of the when-then rules) in order to decide to "respond" by executing the then clause.

From the first category, the used hardware devices manage the system functions of continuous patient monitoring during the exercises and the movement and stimulation of the upper limb that needs to be rehabilitated. The processing and control unit (PCU) determines the correctness of the exercise performed by the patient based on information received from used monitoring devices. The same information is used to update the patient avatar from the virtual environment in which the patient is immersed through the use of VR glasses (HMD). If more than one monitoring device is used, then the system aggregates and synchronizes the gathered information to interpret the status and actions of the patient. The following monitoring devices have been tested: g.tec gUSBamp & gBSanalyze, Kinect V2, Leap Motion, video cam + ArUco markers, Myo armband, EMG, DGTech glove.

The stimulation devices are used to restore and to maintain muscle tone and/or to assist the patient when performing the recovery exercises. The processing and control device synchronize all events and decisions to allow the system to act as a whole. Both hardware equipment and software components are selected to fulfil the system constraints regarding the performance and operational safety. The following stimulation devices have been tested for the best setup and configuration: Oculus Rift/HTC Vive, headphones, Motionstim 8, robotic glove, wireless sticky vibrating motors.

3   Results and Discussion

The system functions dedicated to the therapist are related to patient configuration (search, add and edit), session configurations (patient profile, session content and length, selecting the devices used by each exercise and their configuration), session supervision (through graphs that represent the essential parameters regarding the session in real time) and session history [10], Fig. 2a.

The doctors in the TRAVEE project consortium selected the available exercises and included the most common rehabilitation exercises. These include the Finger Flexion-Extension, the Palm Flexion-Extension, the Forearm Flexion-Extension and the Arm Adduction-Abduction movements. For each selected exercise, the therapist must configure the exercise (the number of repetitions, the duration of repetition and body side left or right), add support (Visual Augmentation, Vibrations, FES), and add monitoring devices (BCI, glove, motion sensor, kinect, leap motion). Every option

Fig. 2   Patient, exercise and session configuration/control

Fig. 3 Virtual environment (patient and world view): **a** patient facing the therapist, **b** patient and therapist facing a mirror, **c** serious game

and potential addition is on the bottom of the session configuration page, Fig. 2b and the flowchart is briefly presented in Fig. 2c. The therapist may choose to edit, run or view/analyse a saved session.

The main features of the system are dedicated to the patient—the subject of the upper limb rehabilitation process. Figure 3 shows that the patient is immersed in a Virtual Environment that includes two avatars (both 3D humanoids). One avatar represents the therapist, which demonstrates the movement that the patient needs to try to reproduce in the real environment. The second avatar represents the user from a first-person point of view that mimics the real-life movements of the patient. The patient may face the therapist Fig. 3a, c or sit next to him/her, both facing a mirror like in a dance room.

There are two minimum recommended configurations: the so called BCI-FES and motion sensor configurations to which other devices can be added. The first configuration consists of a 16 channels biosignal amplifier g.USBamp and an 8-channel neurostimulator Motionstim8. The 12/16 acquired EEG signals are collected from the sensorimotor areas according to the 10–20 International System. The number of EEG signals may vary because four channels may be used, in differential mode, to acquire EOG signals to determine whether the patient is paying attention to the virtual therapist. The 256 Hz sampled EEG signals are preprocessed (filtered with 50 Hz notch filter and 8–30 Hz band pass filter), fed to an algorithm to execute Common Spatial Patterns (CSP) [13–15] spatial filtering, and classify the output as left or right hand movement with Linear Discriminant Analysis (LDA) [16]. For CSP and LDA, the class Common Spatial Patterns 2 from BCI MATLAB&Simulink model provided by g.tec have been used together with g.BSanalyze software (g.tec) for offline data analysis.

As for EOG, the 256 Hz sampled signals are filtered with a moving average filter of 128 samples and then fed to a Simulink block that contains a custom developed algorithm for EOG signal processing. The output of the algorithm is the x-y (HEOG—VEOG) gaze normalized coordinates (Fig. 4) and the number of trigonometric quadrants or centre of the image where the patient is looking on the VR glasses. This is needed to determine whether the patient is dozing off or otherwise ignoring the virtual therapist. If so, the system warns the patient to concentrate/focus on the exercise and pay attention to the virtual therapist.

For the BCI-FES configuration to provide VR feedback based on the patient's imagined movement, the system needs to create a set of spatial filters and classifiers. This is done by recording 4 runs of training data with 20 left and 20 right motor imagery trials in random order [13]. Each 8-second trial consists of:

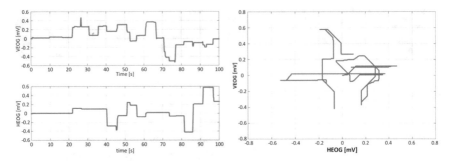

Fig. 4 Eye tracking: **a** HEOG and VEOG, **b** gaze position

(1) A beep at second 2 informing the patient about the upcoming cue;
(2) The cue to perform left or right motor imagery, which is presented from second
     3 until the end (second 8) through both audio (left or right) and video (left or
     right red arrow and left or right therapist hand movement) and indicates that the
     patient needs to start imagining the corresponding movement; and
(3) Visual feedback (in form of the patient avatar moving its hand) begins at second
     4.25. At the same time, the neurostimulator starts to trigger the patient's hand
     movement corresponding to the virtual therapist's cue.

The first two runs are used to build the spatial filters and classifiers. For the
following two runs, each sample classification result is compared with the presented
cue to calculate the error rate for that session as follows:

$$\text{Err} = \left(1 - \left(\frac{\text{Tcc}}{N}\right)\right) \cdot 100$$

where N represents the number of trials and Tcc the number of trials correctly
classified. From the obtained array of 40 error values in the feedback phase, the
mean and minimum error are obtained.

In Fig. 5, an example of output of the LDA classifier can be seen during feedback
phase. Each trial classification output is represented with dotted lines (right-blue,
left-green) and the corresponding average classification output with the solid lines.

Table 1 shows that the mean and minimum classification errors in the feedback
phase are smaller with VR than during the session in which the patient instead
received the visual feedback from a screen. This is because the patient was more
cognitively involved during the exercise using VR, since the VR environment
shielded him from real-world distractions and, in the VR environment, he is no
longer a disabled person. Table 1 contains the mean and minimum classification
errors for seven subjects. The first four subjects (S1–S4) received the visual feedback on a screen in front of them and the following three subjects (S5–S7) received
the visual feedback through VR glasses.

For the second minimum recommended configuration—which use motion
sensors—a device like Kinect (V1/V2), LeapMotion, video cam and ArUko markers,

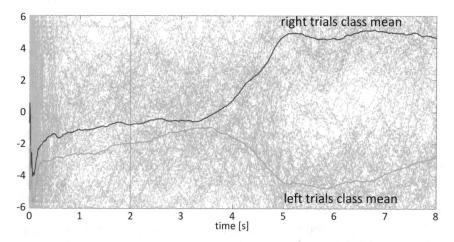

Fig. 5 LDA classification output

Table 1 Mean and minimum classification errors in the feedback phase without (left) and with (right) VR

Subject	Session	Mean Err (%)	Min Err (%)
S1	1	20.62	5.48
	2	22.34	7.11
	3	26.48	19.7
S2	1	23.96	11.97
	2	24.6	14.1
	3	28.83	21.1
S3	1	33.56	22.78
	2	37	21.35
	3	35.58	29.51
S4	1	32.58	24.77
	2	31.54	24.61
	3	37.21	26.22
Mean values	29.53	19.06	
S5	1	18.5	7.36
	2	19.72	10.72
	3	20.8	9.45
S6	1	19.2	6.37
	2	19.25	7.68
	3	19.58	1.95
S7	1	28.19	15
	2	25.53	13.56
	3	21.91	5.13
Mean values	21.41	8.58	

and IMU is used to monitor the patient's hand/arm movement. The neuromotor rehabilitation is divided in three session types: "mirror;" augmented and magnified feedback; and real feedback. All sessions relied on VR glasses to immerse the patient in a virtual environment to receive professional guidance, encouragement, feedback and motivation. The mirror session type was designed to be used immediately after the stroke or brain surgery, when the patient is not able to move the impaired arm or hand. The patient is told to imagine/try/execute the exercise with both arms/hands. The system tracking sensors are set to track only the healthy arm or hand but update both arms/hands of the patient avatar in VR. This way, the patient can see his both arms/hands working just like the therapist instructed, and realizes that s/he can move the hand/arm at will. This visual feedback is very important because it activates the mirror neurons that intermediate learning and closes the causal chain specific to recovery.

If the patient can perform small hand movements of the impaired arm, enough to be detected by the motion sensors, then all executed movements are augmented (session type two). The patient can see a much larger movement than s/he actually executes. The amplification factor decreases from a maximum set value (when the patient's movements are barely detected) to the value of one (when the movement is complete and correctly executed). The augmented feedback transforms the received visual information into knowledge.

The exercises of the third session type should be used after the patient regains partial/total control over the impaired arm and needs motivation to continue therapy by proposing different scenarios and tasks. The difficulties and challenges of the exercises can be adjusted to each patient's condition and progress.

To remind the user about the system and its benefits, each recovery session starts with the therapist and patient sitting in a chair facing each other. The real therapist explains to the patient what s/he will see, hear and must do. Specifically, the virtual therapist will move its left/right hand/arm to demonstrate the exercise to the patient. At the same time, an arrow will appear on the left/right side of the screen, followed by a corresponding left or right audio cue. The patient is instructed to imagine the left or right motor act and do his/her best to execute it. If BCI is used, the patient will receive visual feedback only while s/he is correctly imagining that movement. After this explanation, the VR glasses are mounted and the recovery exercises may begin (see Fig. 6).

Because the prevalence of post stroke spasticity is around 38% [17], some patients with hand spasticity need to perform special exercises to reduce spasticity with a therapist's help [18] before using the TRAVEE. To meet these needs, a second working group on stroke recovery from Technical University of Iasi led by Prof. Poboroniuc designed, build and added a module to TRAVEE to be used especially for despasticisation as well as for recovery exercises for the upper limbs. It consists of a distal exoskeleton glove that can copy the finger movements of the healthy hand by using another glove equipped with bending sensors. It can also actively assist flexion/extension movements of all fingers or each individual finger. The module uses an FES system for better and faster results. This hybrid approach can replace the recovery therapist who usually assists the FES induced movements and can

Fig. 6 Patients using the system

copy the movements of the healthy hand. This mirror-like therapy induces cortical reorganisation and motivates the patient.

The glove (left hand) is made from leather with tendons (metal wires) clamped on the top (dorsal side) and bottom (root) of each finger, as shown in Fig. 7. The right hand is using a textile/leather glove with bending sensor insertion for each finger. Figure 8 presents the hardware architecture of the despasticisation module. The FES module consists of a MotionStim8 neurostimulator that uses two channels for stimulating both the interosseous and extensor digitorum muscles.

The module is not used just to reduce spasticity. It is also integrated in the TRAVEE system, where the therapist can select it as stimulation device and/or as motion sensor based on the type of exercise.

Fig. 7 The distal exoskeleton glove (left hand) and bending sensor glove (right hand)

Fig. 8 The hardware architecture of the despasticisation module

The whole system was first tested on three healthy people. Next, we performed some fine tuning based on their suggestions to improve accuracy and validate system repeatability. Each patient signed an informed consent and an authorization for videos and photographs. The experiments with patients were approved by the institutional review board of the National Institute of Rehabilitation, Physical Medicine and Balneoclimatology from Bucharest, Romania. Patients were women and men with ages between 52 and 79, with post stroke central neuromotor syndrome and stable neurological status, stable consciousness, state, sufficient cognitive functions to allow learning, communication ability, and sufficient physical exercise tolerance.

4 Conclusions

This work presents a BCI-FES system for stroke rehabilitation with the unique combination of BCI and EOG devices to supervise how exercises are performed and monitor patient commitment. The Oculus rift headset increases the patient's immersion in VR. The system must be seen as a software kernel that allows users to define/run a series of rehabilitation exercises using a series of "plugin" devices. By using VR, the patient is not distracted by the real environment and is more cognitively involved during recovery exercises. The patient is focused most of the time, but if s/he loses concentration, the eye tracking system detects this problem and provides a warning. For the BCI-FES configuration, the use of VR makes it possible to provide neurofeedback in one or (rarely) two training sessions.

To our knowledge, the proposed neuromotor recovery system is the only one that includes an eye-tracking device for assessing patient concentration during exercises, enhancing engagement and effectiveness.

Technical performance was validated by testing the system on healthy persons with good knowledge in assistive technologies. The healthy people achieved low control error rates relative to those reported in the literature.

There are two patents pending:

- System, method and software application for automated augmented, gradual and naturalistic representation of human movements 00814/2017, OSIM patent pending.
- Mechatronic glove-neuroprosthesis hybrid system with knitted textile electrodes for hand rehabilitation for patients with neuromotor disabilities 00072/2017, OSIM patent pending.

Acknowledgements This work was supported by the Romanian National Authority for Scientific Research (UEFISCDI), Project 1/2014 Virtual Therapist with Augmented Feedback for Neuromotor Recovery (TRAVEE).

References

1. ENIGMA-Stroke Recovery. http://enigma.ini.usc.edu/ongoing/enigma-stroke-recovery/. Last visit October 2018
2. Johns Hopkins Institute—Strock Centers. www.hopkinsmedicine.org/neurology_neurosurgery/centers_clinics/cerebrovascular/stroke/. Last visit December 2018
3. StrokeBack Project. http://www.strokeback.eu/project.html. Last visit December 2018
4. NIHR—A practical, yet flexible functional electrical stimulation system for upper limb functional rehabilitation, Centres for Health Sciences Research, 2014–2017. https://www.salford.ac.uk/research/health-sciences/research-groups/human-movement-technologies/a-practical,-yet-flexible-functional-electrical-stimulation-system-for-upper-limb-functional-rehabilitation. Last visit December 2018
5. RETRAINER. http://www.retrainer.eu/start/. Last visit December 2018
6. C.M. McCrimmon, C.E. King, P.T. Wang, S.C. Cramer, Z. Nenadic, A.H. Do, Brain-controlled functional electrical stimulation for lower-limb motor recovery in stroke survivors, in *36th Annual International Conference of the IEEE Engineering in Medicine and Biology Society (EMBC)*, 1247–1250, 2014
7. M. Sun, C. Smith, D. Howard, L. Kenney, H. Luckie, K. Waring, P. Taylor, E. Merson, S. Finn, FES-UPP: a flexible functional electrical stimulation system to support upper limb functional activity practice. Front Neurosci. **12**, 449 (2018)
8. O. Ferche, A. Moldoveanu, F. Moldoveanu, The TRAVEE system for neuromotor recovery: Architecture and implementation, in *2017 E-Health and Bioengineering Conference (EHB)*, Sinaia, 2017, pp. 575–578. https://doi.org/10.1109/EHB.2017.7995489
9. S. Caraiman, A. Stan, N. Botezatu, P. Herghelegiu, R.G. Lupu, A. Moldoveanu, Architectural design of a real-time augmented feedback system for neuromotor rehabilitation, in *2015 20th International Conference on Control Systems and Computer Science*, Bucharest, 2015, pp. 850–855. https://doi.org/10.1109/cscs.2015.106
10. R.G. Lupu et al., Virtual reality system for stroke recovery for upper limbs using ArUco markers, in *2017 21st International Conference on System Theory, Control and Computing (ICSTCC)*, Sinaia, 2017, pp. 548–552, https://doi.org/10.1109/icstcc.2017.8107092
11. R.G. Lupu, N. Botezatu, F. Ungureanu, D. Ignat, A. Moldoveanu, Virtual reality based stroke recovery for upper limbs using Leap Motion, in *2016 20th International Conference on System Theory, Control and Computing (ICSTCC)*, Sinaia, 2016, pp. 295–299. https://doi.org/10.1109/icstcc.2016.7790681
12. R.G. Lupu, D.C. Irimia, F. Ungureanu, M.S. Poboroniuc, A. Moldoveanu, BCI and FES based therapy for stroke rehabilitation using VR facilities. Wireless Commun. Mob. Comput. (2018)
13. D.C. Irimia, M.S. Poboroniuc, R. Ortner, B.Z. Allison, C. Guger, Preliminary results of testing a BCI-controlled FES system for post-stroke rehabilitation, in *Proceedings of the 7th Graz Brain-Computer Interface Conference 2017*, September 18th–22nd, Graz, Austria, 2017
14. D.C. Irimia, R. Ortner, M.S. Poboroniuc, B.E. Ignat, C. Guger, High classification accuracy of a motor imagery based brain-computer interface for stroke rehabilitation training. Front. Rob. AI **5**, 130 (2018)
15. S. Lemm, B. Blankertz, T. Dickhaus, K.-R. Müller, Introduction to machine learning for brain imaging. NeuroImage **56**(2), 387–399 (2011)
16. J. Müller-Gerking, G. Pfurtscheller, H. Flyvbjerg, Designing optimal spatial filters for single-trial EEG classification in a movement task. Clin. Neurophysiol. **110**(5), 787–798 (1999)
17. C.L. Watkins, M.J. Leathley, J.M. Gregson, A.P. Moore, T.L. Smith, A.K. Sharma, Prevalence of spasticity post stroke. Clin. Rehab. (2002). https://doi.org/10.1191/0269215502cr512oa
18. D.A. De Silva, N. Venketasubramanian, A. Jr. Roxas, L.P. Kee, Y. Lampl, Understanding Stroke—A Guide for Stroke Survivors and Their Families, 2014. http://www.moleac.com/ebook/Understanding_Stroke_-_Guide_for_Stroke_Survivors.pdf

A Dynamic Window SSVEP-Based Brain-Computer Interface System Using a Spatio-temporal Equalizer

Chen Yang, Xiang Li, Nanlin Shi, Yijun Wang, and Xiaorong Gao

Abstract In this study, we developed a high-speed steady-state visual evoked potential (SSVEP)-based brain-computer interface (BCI) system to address two long-standing challenges in BCIs: tedious user training and low applicability for target users. We designed a training-free method with low computational complexity called the spatio-temporal equalization dynamic window (STE-DW) recognition algorithm. The algorithm uses the adaptive spatio-temporal equalizer to equalize the signal from both the spatial and temporal domains to reduce the adverse effects of colored noise. We then implemented this algorithm into a dual-platform distributed system to facilitate BCI spelling applications. Finally, the complete system was validated by both healthy users and one person with Amyotrophic Lateral Sclerosis (ALS). The results suggest that the STE-DW algorithm integrated BCI system is a robust and easy-to-use system, which can provide a high-performance, and training-free BCI experience for both healthy and disabled users.

Keywords Brain-Computer Interface (BCI) · Steady-State Visual Evoked Potential (SSVEP) · Amyotrophic Lateral Sclerosis (ALS) · Spatio-temporal equalization · Dynamic window

C. Yang · X. Li · N. Shi · X. Gao (✉)
Department of Biomedical Engineering, Tsinghua University, Beijing, People's Republic of China
e-mail: gxr-dea@mail.tsinghua.edu.cn

C. Yang
School of Electronic Engineering, Beijing University of Posts and Telecommunications, Beijing, People's Republic of China

Y. Wang
Institute of Semiconductors, Chinese Academy of Sciences, Beijing, People's Republic of China

© The Author(s), under exclusive license to Springer Nature Switzerland AG 2020   87
C. Guger et al. (eds.), *Brain–Computer Interface Research*,
SpringerBriefs in Electrical and Computer Engineering,
https://doi.org/10.1007/978-3-030-49583-1_9

1   Introduction

In recent years, brain-computer interfaces (BCIs) have attracted attention from researchers in the field of neural engineering, neuroscience, and clinical rehabilitation (BCIs) as a new direct communication pathway for individuals with severe neuromuscular disorders [1–3]. The steady-state visual evoked potential (SSVEP)-based BCI, one kind of the most promising types of BCIs, has attracted more and more attention due to its advantages such as high information transfer rate (ITR), excellent signal-to-noise ratio (SNR), easy quantification, and minimal user training. SSVEP is a periodic electrophysiological signal evoked by visual stimulus at a fixed frequency and is most prominent in the visual cortex. Therefore, SSVEP can be used as the primary feature for gaze or attention detection in a BCI system. In traditional SSVEP-BCIs, each target flickers at a specific frequency, and a target can be determined by identifying corresponding frequencies in the elicited SSVEP.

Currently, two major problems remain in the traditional SSVEP recognition algorithms. First, most of the traditional SSVEP recognition algorithms adopt a spatial filtering process based on the assumption that the different background noises in the time domain are independent and identically distributed, and the power spectral density in the frequency domain is uniformly distributed [4, 5]. However, many studies have pointed out that the background noise of EEG is not white, but a colored noise, whose power spectrum shows 1/f distribution [6, 7]. Second, traditional SSVEP recognition algorithms mainly use the "fixed window length" in the time domain [4, 5, 8], leading to a constant recognition time for each trial. These algorithms generally estimate the optimal window length to optimize BCI performance. However, due to the non-stationary nature of EEG, a unified window length is not optimal across trials and subjects.

To solve these problems, based on the theory of adaptive equalization and hypothesis testing, this study designed a 40-target SSVEP-based BCI using the spatio-temporal equalization and dynamic window (STE-DW) recognition algorithm. Specifically, a spatio-temporal equalization algorithm is used to reduce the adverse effects of space-time correlation of background noise. Also, based on the theory of multiple hypotheses testing, a stimulus termination criterion is used to implement dynamic window control. The proposed system can adaptively equalize the signal from both the spatial and temporal domains to reduce the adverse effects of colored noise and dynamically determine the stimulus window length in real-time with the recognition process.

Despite the extensive effort devoted to developing high-performance BCIs, studies seldom involve actual Amyotrophic Lateral Sclerosis (ALS) patients in complete locked-in state (CLIS), who are necessary to use and validate BCIs. Several studies applied BCIs to real patients, predominantly using the P300 paradigm. For example, Donchin et al. performed preliminary tests on 3 ALS patients and showed promising results [9]. Similar studies including Silvoni et al. [10] and Wolpaw et al. [11] conducted on a relatively large population (more than 20) have shown the feasibility of non-invasive P300 BCIs. Vansteensel et al. [12] even performed invasive experiments on ALS patients, during which they inserted electrodes on the cerebral cortex and reported that, after extensive training, patients attained precise control over long-term use. However, both non-invasive and invasive P300 BCIs may exhaust ALS patients and entail other problems. To evaluate the proposed system and help address the problem of inadequate involvement of ALS patients, we presented a case study with an ALS patient in a locked-in state using our system.

This paper is organized as follows. After the introduction, Sect. 2 presents further information about the dynamic window method and spatio-temporal equalizer. Section 3 reviews our system implementation, in which we explain the software system configuration in detail. Section 4 describes the experiment design and the result with healthy subjects. Section 5 describes the application of the proposed BCI system to an ALS patient and reported his feedback. Finally, we conclude this chapter in Sect. 6.

2 Methodology

2.1 Theoretical Model

2.1.1 SSVEP-EEG Data Model

The EEG evoked model, which is shown in Fig. 1, can be expressed as:

$$x(n) = s(n) + w(n), n = 1, \dots, N \tag{1}$$

where N represents the time sampling points, $x(n)$ represents the L channel EEG data at time n, and the sampling time is N; $s(n)$ represents the SSVEP component; and $w(n)$ denotes the background noise, including spontaneous EEG and system noise. It is generally agreed that the SSVEP component $s(n)$ is a linear mixture of multi-frequency sinusoidal signals, and is independent of background noise [13]. $s(n)$ can be expressed as:

Fig. 1 The conduction and equalization model of EEG background noise

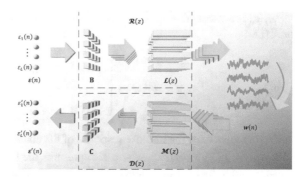

$$s(n) = \mathbf{A}\boldsymbol{\phi}(n) \tag{2}$$

$$\boldsymbol{\phi}(n) = \frac{1}{\sqrt{N}}\left[e^{-j\omega n}\ e^{-j\omega n}\ \cdots\ e^{-jl\omega n}\ e^{-jl\omega n}\right]^{\mathrm{T}}$$

where $\mathbf{A} \in \mathbb{R}^{L \times l}$ denotes the aliasing matrix of the sinusoidal template signal. $\boldsymbol{\phi}(n) \in \mathbb{R}^{l \times 1}$ denotes l-dimensional, complex sinusoidal template signal, which is a semi-orthogonal matrix.

Background noise $\boldsymbol{w}(n)$ has three characteristics:

(1) Background noise has a short-term stationary feature.
(2) Background noise can be regarded as a zero-mean Gaussian random signal.
(3) Background noise has spatial and temporal correlation.

In this study, a channel system based on Finite Impulse Response—Multiple Input Multiple Output (FIR-MIMO) is proposed to simulate the generation and conduction process of the background EEG according to its characteristics. The process is described as follows:

(1) First, L independent and identically distributed Gaussian sources produce white noise $\boldsymbol{\varepsilon}(n) = \left[\varepsilon_1(n) \cdots \varepsilon_L(n)\right]^{\mathrm{T}}$.
(2) Then, $\boldsymbol{\varepsilon}(n)$ passes through the non-stationary zero-order channel $\mathbf{B}$, and forms non-stationary noise with spatial correlation.
(3) Finally, the noise passes through the stationary high-order channel $\mathcal{L}(z)$ again and forms noise that has both spatial and temporal correlation.

In this process, $\mathbf{B}$ is non-stationary and is used to simulate the source signal aliasing process; $\mathcal{L}(z)$ is stationary within a period of time and simulates the conductive process of the background EEG transmitting from the cortex to the scalp. Specifically, $\mathbf{B}$ is defined as a $L \times L$ full rank square matrix, and $\mathcal{L}(z)$ is an equalizable linear system, which satisfies $\text{rank}(\mathcal{L}(z)) = L$ for all non-zero $z \in \mathbf{C}^2$ [14]. Thus, the entire transmission process can be described by the high order FIR-MIMO channel $\mathcal{R}(z)$,

$$\mathcal{R}(z) = \mathcal{L}(z)\mathbf{B} \tag{3}$$

$\gamma_{ij}(z)$ represents the transfer channel of the j-th noise source to the i-th receiver in $\mathcal{R}(z)$,

$$\mathcal{R}(z) = \begin{bmatrix} \gamma_{11}(z) & \cdots & \gamma_{1L}(z) \\ \vdots & \ddots & \vdots \\ \gamma_{L1}(z) & \cdots & \gamma_{LL}(z) \end{bmatrix} \tag{4}$$

and $r_{ij}(n)$ represents the impulse response of the channel $\gamma_{ij}(z)$. It can be expressed as $\gamma_{ij}(z) = \sum_{n=0}^{K} r_{ij}(n)z^{-n}$, where K represents the highest order of $\mathcal{R}(z)$. So, the noise recorded by the i-th channel of EEG receiver can be expressed as:

$$w_i(n) = \sum_{j=1}^{L} r_{ij}(n) * \varepsilon_j(n) = \sum_{j=1}^{L} \sum_{\tau=0}^{T} r_{ij}(\tau)\varepsilon_j(n - \tau) \tag{5}$$

where $*$ denotes a linear convolution and $\boldsymbol{w}(n) = \begin{bmatrix} w_1(n) & w_2(n) & \cdots & w_L(n) \end{bmatrix}^{\mathrm{T}}$.

2.1.2 Parameter Estimation

According to the model assumptions, the background noise transmission process can be described using FIR-MIMO channel model $\mathcal{R}(z)$. There is always an equalizer $\mathcal{D}(z)$ that can filter the background noise $\boldsymbol{w}(n)$ to the equalized noise $\boldsymbol{\varepsilon}'(n)$ without the spatial and temporal correlation. Figure 1 shows the background noise conduction and equalization model.

According to (3), the matrix product of $\mathcal{R}(z)$ is concatenated by the non-stationary channel $\mathbf{B}$ and stationary channel $\mathcal{L}(z)$. Thus, the non-stationary equalizer $\mathbf{C}$ and the stationary equalizer $\mathcal{M}(z)$ are used to equalize the non-stationary channel $\mathbf{B}$ and the stationary channel $\mathcal{L}(z)$, respectively. $\mathcal{D}(z)$ can be expressed as:

$$\mathcal{D}(z) = \mathbf{C}\mathcal{M}(z) \tag{6}$$

where $\mathcal{M}(z)$ is a ρ-orders stationary FIR spatio-temporal equalizer, and $\mathbf{C}$ is a zero-order non-stationary spatial equalizer. $\mathcal{M}(z)$ and $\mathbf{C}$ satisfy that:

$$\mathcal{M}(z)\mathcal{L}(z)\mathcal{L}(1/z^*)^{\mathrm{H}}\mathcal{M}(1/z^*)^{\mathrm{H}} = \mathbf{I}_L \tag{7}$$

$$\mathbf{CBB}^{\mathrm{H}}\mathbf{C}^{\mathrm{H}} = \mathbf{I}_L \tag{8}$$

That is, the product of $\mathcal{M}(z)$ and $\mathcal{L}(z)$ is a paraunitary matrix, and the product of $\mathbf{C}$ and $\mathbf{B}$ is a unitary matrix. Let $\mathcal{F}_N(\mathcal{D})$ represent the transfer matrix of $\mathcal{D}(z)$, and $\mathcal{D}_k$ represents the k-th order transfer coefficient of $\mathcal{D}(z)$. That is,

$$\mathcal{F}_N(\mathcal{D}) = \left.\begin{bmatrix} \mathcal{D}_0 & & & & \\ \vdots & \mathcal{D}_0 & & & \\ \mathcal{D}_\rho & \vdots & \ddots & & \\ & \mathcal{D}_\rho & & \mathcal{D}_0 & \\ & & \ddots & \vdots & \ddots \\ & & & \mathcal{D}_\rho & \cdots & \mathcal{D}_0 \end{bmatrix}\right\} N\ blocks \tag{9}$$

$$\mathcal{D}(z) = \sum_{k=0}^{\rho} \mathcal{D}_k z^{-k} \tag{10}$$

The background noise can be equalized by $\mathcal{F}_N(\mathcal{D})$, so $\mathcal{F}_N(\mathcal{D})$ can be regarded as an estimate of $\Sigma_w^{-1/2}$. That is,

$$\Sigma_w^{-1/2} = \mathcal{F}_N(\mathcal{D}) \tag{11}$$

That is,

$$\text{vec}\left(\hat{\mathbf{A}}\right) = \left[\left(\mathbf{\Phi}^H \otimes \mathbf{I}_L\right)^H \mathcal{F}_N(\mathcal{D})^H \mathcal{F}_N(\mathcal{D})\left(\mathbf{\Phi}^H \otimes \mathbf{I}_L\right)\right]^{-1}$$
$$\left(\mathbf{\Phi}^H \otimes \mathbf{I}_L\right)^H \mathcal{F}_N(\mathcal{D})^H \mathcal{F}_N(\mathcal{D})\bar{x} \tag{12}$$

$$\bar{x} = \left[x(1)^T \cdots x(N)^T\right]^T$$

$$\mathbf{\Phi} = \left[\boldsymbol{\phi}(1) \cdots \boldsymbol{\phi}(N)\right]$$

where vec() denotes a vectorization and $\otimes$ denotes a kronecker product.

2.2 Dynamic Window Hypothesis

According to (1) and (2), for each SSVEP-BCI system with multiple targets, detection of each target can be abstracted as a hypothesis:

$$\mathcal{H}^{\{q\}}x(n) = s^{\{q\}}(n) + w(n) \tag{13}$$

where $\mathcal{H}^{\{q\}}$ represents the EEG data $x(n)$ that contains the SSVEP component $s^{\{q\}}(n)$, which is evoked by the q-th stimulus frequency.

The probability density function under the hypothesis $\mathcal{H}^{\{q\}}$ can be expressed as:

$$p\left(\mathbf{X}|\mathcal{H}^{\{q\}}\right) = \frac{1}{(2\pi)^{\frac{LN}{2}} \det(\Sigma_w)^{1/2}} e^{-\frac{1}{2}\hat{\theta}^{\{q\}H}\hat{\theta}^{\{q\}}} \tag{14}$$

$$\hat{\theta}^{\{q\}} = \Sigma_w^{-1/2}\text{vec}(\mathbf{X}) - \Sigma_w^{-1/2}\left(\mathbf{\Phi}^{\{q\}H} \otimes \mathbf{I}_L\right)\text{vec}\left(\mathbf{A}^{\{q\}}\right) \tag{15}$$

Suppose that the SSVEP-BCI system contains Q targets (the range of q can be defined as $1, \ldots, Q$), and the EEG data are continuously received. Then, an additional "erasure decision" [15] $\mathcal{H}^{\{0\}}$ can be added to hypothesize. When $\mathcal{H}^{\{0\}}$ is decided, it indicates that the current data are not enough to make a reasonable decision, so the system needs to wait for more new data [16]. It can be expressed as:

$$
\begin{cases}
\mathcal{H}^{\{0\}} : \text{reject} \\
\mathcal{H}^{\{1\}} : x(n) = s^{\{1\}}(n) + w(n) \\
\quad\quad\quad \vdots \\
\mathcal{H}^{\{Q\}} : x(n) = s^{\{Q\}}(n) + w(n)
\end{cases} \tag{16}
$$

Under the condition of $\mathcal{H}^{\{1\}} \cdots \mathcal{H}^{\{Q\}}$, the probability of each target is $1/Q$. $\mathcal{H}^{\{q_{\mathcal{H}}\}}$ represents the $q_{\mathcal{H}}$-th hypothesis is selected, $c(\mathcal{H}^{\{q_{\mathcal{H}}\}}, \mathcal{H}^{\{q\}})$ indicates the cost of the situation where the real case is $\mathcal{H}^{\{q\}}$, but the system makes the decision of $\mathcal{H}^{\{q_{\mathcal{H}}\}}$.

$$
\begin{cases}
c(\mathcal{H}^{\{q_{\mathcal{H}}\}}, \mathcal{H}^{\{q\}}) = 0 \; q = q_{\mathcal{H}}, q_{\mathcal{H}} \neq 0 \\
c(\mathcal{H}^{\{q_{\mathcal{H}}\}}, \mathcal{H}^{\{q\}}) = 1 \; q \neq q_{\mathcal{H}}, q_{\mathcal{H}} \neq 0 \\
c(\mathcal{H}^{\{q_{\mathcal{H}}\}}, \mathcal{H}^{\{q\}}) = \epsilon \; q_{\mathcal{H}} = 0
\end{cases}
$$

The statistics γ can be defined as

$$
\gamma = \min_{q_{\mathcal{H}}} \left[1 - \frac{p(x(n)|\mathcal{H}^{\{q_{\mathcal{H}}\}})}{\sum_{q=1}^{Q} p(x(n)|\mathcal{H}^{\{q\}})} \right] \tag{17}
$$

where $p(x(n)|\mathcal{H}^{\{q\}})$ represents the conditional probability of $x(n)$ in the case of $\mathcal{H}^{\{q\}}$. It can be calculated by (12). Usually, as the stimulation time increases, the statistic γ will gradually decrease.

When $\gamma \geq \epsilon$, the cost of "erasure decisions" $\mathcal{H}^{\{0\}}$ is considered as the smallest of all hypotheses, which indicates that the system should continue to receive data. When $\gamma < \epsilon$, it suggests that at least one cost of the hypothesizes $\mathcal{H}^{\{1\}} \cdots \mathcal{H}^{\{Q\}}$ is lower than the cost of $\mathcal{H}^{\{0\}}$, so the system can stop data acquisition. The system then proceeds to give the recognition result by the specific decision criteria. It should be noted that in this study the dynamic window hypothesis is only used for selecting dynamic window length to determine if the termination acquisition condition is met.

3 System Implementation

3.1 System Framework

This system uses a dual-platform distributed system-based architecture design to meet the real-time stimulation and processing requirements for the BCI application, as shown in Fig. 2. One of these platforms is the stimulation platform, which is mainly responsible for stimulation-related tasks such as target generation, display control, recognition, and feedback. The other is the data processing platform, which is responsible for the reception, analysis, and peripheral control of EEG signals. The two platforms communicate via TCP/IP to jointly implement the stimulation, processing, and feedback functions of the BCI system.

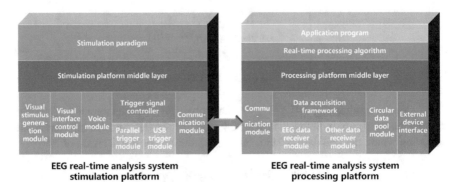

Fig. 2 The architecture of a distributed system using the spatio-temporal equalization and dynamic window (STE-DW) recognition algorithm

3.1.1 Stimulation Platform Framework

The stimulation platform can support the development of stimulation paradigms in BCI systems. It provides a series of interfaces and functional modules to meet the needs of different stimulation paradigms. If researchers wish to develop new BCI applications, they can call all the functional modules provided by the platform framework to improve development efficiency.

This platform is usually divided into three layers from the bottom to top: the functional module layer, the platform middle layer, and the stimulation paradigm layer.

(1) **Functional module layer**: The functional module layer mainly includes specific implementation modules for various functions.

   a. **Visual stimulus generation module**: The function of this module is to generate and save images of visual stimulation examples. The stimulation platform uses a pre-load method to generate stimulation paradigms to ensure the stability of the stimulation. By invoking the visual stimulus generation module, the BCI application can quickly generate the required paradigm stimulus and store and pre-load the stimulus paradigm properly.

   b. **Visual interface control module**: The visual interface control module can control the flow of the stimulation paradigm. In order to improve the generality, the module itself does not include the specific implementation process but can provide many display control functions, and realize the procedural control of the experimental paradigm through the paradigm process control script.

   c. **Voice module**: The voice module can provide users with voice feedback, and let users receive feedback content of the voice reading.

   d. **Trigger signal controller module**: This module is compatible with the parallel port, USB interface, and other trigger signal sending methods. This module will unify the interface and select the corresponding trigger sending method according to the configuration file. Therefore, during the development of the upper-level stimulation paradigm, it can be supported on multiple trigger modes and multiple types of EEG amplifiers by calling standard interfaces.

   e. **Communication module:** The communication module establishes a data transmission path between the stimulation platform and the signal processing platform. Therefore, the messages from the stimulation platform can be quickly transmitted to the signal processing platform, and the stimulus platform can receive feedback from the signal processing platform in real time.

(2) **Stimulation platform middle layer**: The stimulation platform middle layer divides the application control layer and the functional module layer. It shields the execution details of each functional module for specific applications, thereby improving the compatibility of BCI applications for different hardware platforms. At the same time, it provides a unified, standardized interface for different functional modules and also enables functional modules to be reused by different BCI applications.

(3) **Stimulation paradigm layer**: The stimulation paradigm layer is primarily responsible for the paradigm generation and process control of specific applications. This module contains the specific paradigm stimulus generation script and process control script. The visual stimulus generation module can be called through the middle layer of the platform to generate the required pre-loaded stimulation sequence. Moreover, through the middle layer of the platform, calling the visual interface control module can realize the functions of target prompting, starting and stopping stimulation, and feedback presentation.

3.1.2 Processing Platform Framework

This platform, including data transmission, data buffering and results sending module, mostly performs as the data processing unit for the BCI system, which can fulfill the requirements of various BCI applications. The processing platform can be generally divided into four structural layers, which are functional module layer, middle platform layer, data analysis layer, and application control layer from bottom to the top order.

(1) **Functional module layer**: As the same as the stimulation platform, the functional module layer in the processing platform also contains a various implementation of concrete functions.

   a. **Data acquisition framework**: EEG data can be collected in real time through this framework. The data access layer framework uses a unified and standardized interface to achieve standardized reception of data from multiple sources. The framework can eliminate systematic differences of different hardware, and provide a unified data flow control interface and standardized data structure for upper-layer applications. Therefore, there is no need to consider the details of different hardware when developing upper-layer applications.

At the same time, the access layer framework also supports multi-modal data, such as multiple inputs of EEG data, EOG data, and eye movement data. This framework means that the system can expand the function modules of data collection at any time as needed.

b. **Circular data pool module**: The circular data pool module can be used as a real-time data buffer to isolate data collection and data processing algorithms asynchronously so that the recognition and analysis algorithms with different computing speed can run correctly under this platform.

c. **External device interface module**: The external device interface module is mainly used to connect external systems or external equipment and can be expanded as needed.

d. **Communication module**: As same as the communication module in the stimulation platform, this module can receive message instructions from the stimulation platform in real time and send feedback of the recognition results to the stimulation platform in time.

(2) **Processing platform middle layer**: The Processing platform middle layer has the same function as the stimulation platform middle layer.

(3) **Real-time processing algorithm layer**: The real-time processing algorithm layer can process a variety of data signals in real time and transmit the recognition results to the application layer. The data processing layer mainly provides recognition algorithms for EEG signals. Besides, users need to strictly control the computational complexity of the algorithm to meet the needs of real-time processing.

(4) **Application program layer**: This layer module is mainly responsible for the execution of the specific application processing program, including the start/stop of the EEG processing algorithm, and the feedback of the EEG recognition results.

In this system, the application layer can convert the frequency and phase information identified by the real-time processing algorithm layer into corresponding text symbols and feed them back to the stimulation platform through the communication module.

3.2 System Logic Structure

This system adopted a distributed design, meaning that the simulator, collector, and operator are deployed on different hardware devices. The logical structure of the online system is shown in Fig. 3. Considering the computational and transmission lag

Fig. 3 The logical structure
of the online system

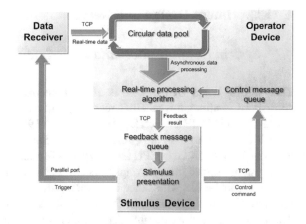

between devices, the systems across three hardware platforms all inherit a telecommunication buffer mechanism. In this design, the stimulator, which is responsible for the presenting paradigm, connects to the operator through TCP.

This protocol enables the stimulator to receive and process the instructions, including feedback in real-time from the operator. The stimulus presentation from the stimulus imposes quite high requirements for real-time processing, which usually occupies the entire central processing unit (CPU). Therefore, instruction and feedback processing may be delayed. To address this shortcoming, the proposed system is equipped with a feedback messages queue in the stimulator to buffer the instructions of the messages from the operator and separate the operator feedback and stimulation time lag. Furthermore, the stimulator and collector are connected through a parallel port or USB port, which enables the system to record trigger signals in millisecond precision.

For acquisition, EEG signals are acquired by the collector, which also transports data by TCP protocols. The collector sends real-time data to the operator in a fixed format. The durations of data transportation, mostly less than 50 ms, vary with the type of collector.

The data operator receives the data package transmitted by the collector and calls algorithms for real-time calculations. The computing speed of operators is different according to distinctive types of detection algorithms. To address the mismatch between the speed of transmissions and receptions, the operator used a circular data pool to save the received data package. The data pool can quickly receive the report of data transported from the collector. The whole processing system can select data of the required length for analysis. Similar to the stimulator, a control message queue was introduced in the operator for receiving control instructions that came from the stimulator.

4 Experiment Design and Results

4.1 System Configuration and Experimental Paradigm

This SSVEP-based BCI system is a real-time online processing system, which consists of a stimulator, an EEG recorder, and an operator, as shown in Fig. 4a. An ASUS VG278HE 27-inch LCD monitor with 1920 × 1080 resolution and 60 Hz refresh rate was used as the stimulator and a Surface Pro 4 tablet PC with dual-core i5-6300U@2.4 GHz CPU and DDR3 4 GB memory was used as the operator. This system utilized the SynAmps2 system (Neuroscan, Inc.) as the EEG recorder, which is compatible with other amplifiers. All of the programs were developed under MATLAB 2015b using the Psychophysics Toolbox Version 333 (PTB-3) [17]. Throughout the whole experiment, the stimulator presented multi-target stimulation to the subject and sent a start trigger and an end trigger signal to the EEG collector at the beginning and end of each trial, respectively. Then, the operator received EEG data from the EEG amplifier in real time, sent a stop trigger to the stimulator when the stimulus stop condition was satisfied, and sent the feedback to the stimulus device after completing the recognition. In order to facilitate the comparison of algorithm performance and eliminate the interference from experimental conditions, this experiment adopts a paradigm design (the frequency, initial phase, position, and size of each target block) from [18]. A 5 × 8 stimulation matrix containing 40 characters

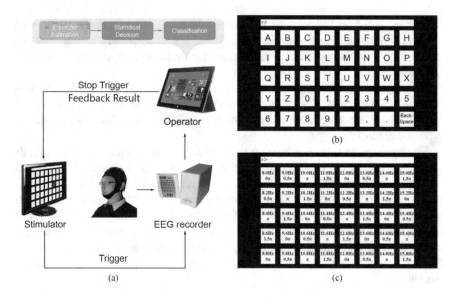

Fig. 4 **a** Experimental system configuration, in which the red box in the display indicates the target cue, **b** Experimental stimulation interface, **c** The frequency and the initial phase of stimulus targets (the first line indicates the target frequency and the second line indicates the initial phase in each rectangle)

(A-Z total of 26 English letters, 0–9 total of 10 digits, three other symbols, and the backspace) was presented on the stimulator, as shown in Fig. 4b. The frequency and the initial phase values of all stimuli are shown in Fig. 4c.

4.2 Data Acquisition

EEG data were acquired using a 64-channel EEG cap at a sampling rate of 1000 Hz. During the experiment, the raw data were preprocessed by a 1–100 Hz band-pass filter and a 50 Hz notch filter. Then, all data were down-sampled to 250 Hz. Only nine channels of data (Pz, PO5, PO3, POz, PO4, PO6, O1, Oz, and O2) were used to record SSVEPs. The reference electrode was located at the vertex. All electrodes were placed according to the international 10–20 system. Electrode impedances were kept below 10 kΩ. Seventeen healthy subjects participated in this study, including 12 young subjects aged 17 to 30 years old (5 females) and 5 old subjects aged 59 to 68 years old (2 females). One subject (male, aged 66 years old) was excluded because his body could not support him in completing the experiments. Each subject signed his or her written informed consent before the experiment and was paid for participation. This study was approved by the Research Ethics Committee of Tsinghua University.

4.3 Online Experiment

The online experiment was designed to verify the practical performance of the STE-DW algorithm. The filter bank canonical correlation analysis (FBCCA) algorithm was used for performance comparison. Sixteen subjects participated in the online experiment. The online experiment consisted of 4 blocks, each of which contained 80 trials. Each target was triggered once in both the first 40 trials and the last 40 trials with a random order. At the beginning 0.5 s of each trial, which is called the cue stage, a red block appeared on the screen to indicate the target position (as shown in stimulator in Fig. 4a). Subjects need to turn their attention quickly to the cued target. During the stimulation process, all stimuli flickered simultaneously at their predetermined frequencies. The subjects were asked to focus on the prompted target. Meanwhile, to facilitate visual fixation, a red triangle appeared below the prompt target. During the online experiment, the EEG amplifier sent a data packet to the operator every 40 μs. The STE-DW algorithm dynamically determined the stimulus duration based on the result of the hypothesis test. The stimulus duration of the FBCCA algorithm was set to 1.25 s. Within 0.5 s after the stimulus, the recognition result would be fed back in the form of characters typed in the text input field at the top of the screen.

4.4 Results

Figure 5 shows the results of the online experiment. FBCCA used a 1.25 s fixed stimulus duration according to the optimization results given in [19], while the STE-DW algorithm used a dynamic time window with the parameters optimized from a public dataset [18]. The results show that the STE-DW algorithm led to a higher ITR (average: 113.4 bits/min) for most of the subjects compared with FBCCA (average: 90.7 bits/min), and the accuracy of all subjects was higher than 70%. In contrast, nine subjects obtained accuracy below 70% using FBCCA. For subjects with a low SNR, the STE-DW algorithm can achieve higher recognition accuracy by dynamically prolonging the stimulation time. For subjects with high SNR, the STE-DW algorithm can shorten the stimulation time to improve the ITRs.

Figure 6 illustrates that, except for trial time, other performance measures of the STE-DW algorithm are better than those of the FBCCA algorithm. During practical use, the STE-DW algorithm finishes the analysis step in 1 ms, and the adaptive updating of the stationary equalizer can be completed within 10 μs.

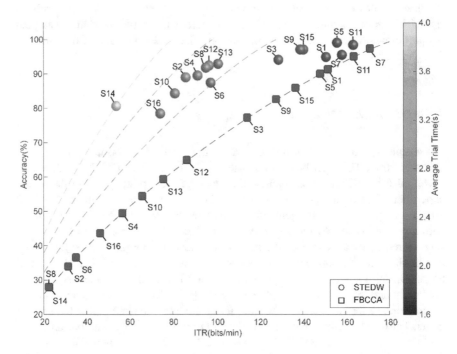

Fig. 5 The average ITR, accuracy, and trial duration for each subject using the STE-DW algorithm and the FBCCA algorithm in an online experiment. The dotted lines from bottom to top showed the fitting relation between accuracy and ITR at the different trial time (T = 1.75 s, 2.25 s, 3.25 s, 4 s) respectively

Fig. 6 Performance comparison of the STE-DW and FBCCA algorithm in the online experiment

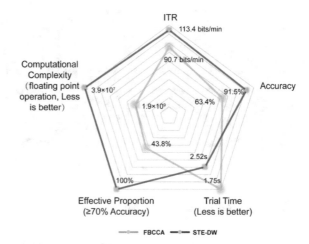

5   Evaluation with ALS Patient

5.1   Patient Information

The patient who participated in this study is a 38 year old male and has been diagnosed ALS for 11 years, as shown in Table 1. Because of the loss of his motor abilities, he can no longer initiate voluntary communication through speech or body gestures. However, his cognitive abilities are good; he has been writing, spelling, and expressing himself mainly via an eye-tracker, together with facial expression (i.e., twitches using the only remaining eyebrow muscles).

His essential life support mainly relies on respiratory and stomach intubation. This experiment was approved by the Institution Review Board of Tsinghua University, and the patient provided informed consent.

Table 1 Patient information

ALS1	
Gender	Male
Age	37
Time since diagnosis	11 years
Artificial ventilation	Yes
Limb muscle control	Absent
Eye movement	Weak
Eyelid muscle control	Not available
Communication mode	Eye movement

Fig. 7 BCI systems experiments in home and studio environments

5.2 Implementation Detail

The system implementations are already specified in previous sections. With the aforementioned system, we utilized a 9-channel electrode cap (Pz, PO5, PO3, POz, PO4, PO6, O1, Oz, and O2) and a wireless EEG amplifier (Neuracle) to acquire the SSVEP signals from the occipital region.

Besides the hardware implementation, we specially designed a double-spelling strategy to fulfill Chinese spelling tasks for patient's convenience. The double-spelling strategy was based on Chinese syllabification, where the user was spelling a whole Chinese character by first selecting the initial target (constant) on the first page and then selecting the respective second target on the next page. This spelling strategy is considered efficient because it costs exactly two steps to spell a character, which is the minimum compared with other strategies like the QuanPin input method.

The proposed system with the double-spelling strategy was tested under the indoor ambient environment and a TV show studio with strong electromagnetic noise sources (Fig. 7). The system worked well in these different environments, and the spelling results were satisfying for both the patient and the person he communicated with.

5.3 Online Results

Online spelling texts were collected and analyzed selectively. We list the representative texts spelled by the patient in the format of spelling contents, actual input, total number of inputs, number of correct inputs, and total time. The patient managed to spell 363 of total 406 Chinese characters correctly, which equals 89.4% accuracy at the character level. Compared to the benchmark level, which is commonly considered to be 70%, the proposed system can be an accurate and effective spelling tool. In addition to the high accuracy, the corresponding ITR of 22.2 bits/min is the highest performance ever reported in BCI studies with ALS patients, to the best of our knowledge.

Furthermore, we should emphasize that the system does not burden the patient with an excessive training process. Comparing to a non-invasive BCI spelling system that typically can spell two English characters per minute after weeks of training, the SSVEP speller allows spelling of the equivalent of 1.5–2.5 Chinese characters per minute without any training procedure.

5.4 Patient Feedback

The patient was able to type greetings after a short introduction and a 30-second calibration, which suggested that our system was easy to use at the first try. Over the following months, the patient practiced once a week, and soon became a veteran in BCI spelling and succeeded in typing a random-picked Chinese poem in a TV show. In addition, the online spelling system was frequently used in daily life, where the patient expressed his personal feelings, care requirements, and suggestions to improve the present system.

The patient felt satisfied with the BCI system, which can be indicated from the contents in Table 2. The patient spelled that he does not think the system is complicated for him (in Chinese, "不复杂我都记住啦" after we explained how to use the speller.

During the experiment, we specifically worried that he might develop visual fatigue, yet the patient responded that he did not feel tired (in Chinese, "不累"), and even suggested that we should increase the experiment time and intensity (in Chinese "可以增强度"). From these concrete responses, we confirm that the patient generally feels satisfied with the spelling system. The spelling experience provided by our system is natural and efficient for him. The patient also claimed that what he needs the most is the Microsoft Word-like applications, which would be convenient for him to record thoughts and the articles he has been trying to write. ("我想用它每天写点东西，能不能有个沃得文档" in Chinese). Thus, the next step in system development should consider patient needs such as these.

6 Conclusion

In conclusion, by establishing a mathematical model, the STE-DW algorithm was applied to implement a BCI system using dynamic time window detection based on the hypothesis test. The system has simple design parameters, low computational complexity, and does not require individual training data. Compared with traditional training-free algorithms, it can effectively improve ITR and recognition accuracy. At the same time, because of the adaptive window detection mechanism, the BCI algorithm can achieve high adaptability and may be applied to the vast majority of subjects, including ALS patients. Therefore, our system based on the STE-DW algorithm has achieved a good balance in terms of efficiency, effectiveness, economy

Table 2 Spelling content analysis

Spelling Content	Actual Input	Total Number of Inputs	Number of Correct Inputs	Total Spelling Times(s)
高兴就好 可以增强度	g 0 g ao x ing 2 j iu h ao 3 2 k e y 3 z v # o # # z ai # eng j ia 2 q # # q iang d u # 6	13 26	12 20	123 311.3
爸爸速干衣 我相写点设计思想用遮盖你 们先回血性留下姐姐用笔记 写	b ab 2 s u g # g 8 0 9 # g an 3 y i f 5 # w o x 5 iang 3 x 5 iang 3 x ie d ian 5 2 sh e j i 2 si x iang 2 y ong 1 2 b # zh e g 4 k # n ing # i m 4 x ian 3 h ui 3 d # x ue x iao b # # 7 1 o # i # iu x ia l ai 2 # # j # j ie j ie 2 y ong zh 6 # b i j i 6 x ie x ia l uan # 4 e b	21 99	19 88	232.7 1342
不累 天命风流你	t # b u l ei 4 4 t ian m ing r 3 # f eng l iu 3 3 n i b u # # 2	8 20	7 17	71.2 254.9
我想用它每天写点东西，能 不能有个沃得文档	h # w o x 5 5 z # # 4 # x iang 1 x # 2 y ong 2 t 4 m ei t 8 x ie d ian t # 5 2 d ong x i 2, n eng 2 b u 2 n eng 2 y ou # ou g 2 w o 5 d e j # 6 w en d ang 2	69	61	600.8
怎么找寻，我可以看见别人 说话的绿框嘛	z en m 2 zh ao x 4, w o 2 k e h # y 3 k an j 3 b ie r 2 sh uo h ua 2 d e 2 l v k uang 6 6 m 9	42	41	589.7
辛苦啦 这个电脑什么牌子雷蛇	x in k 2 1 3 zh e g 2 d ian n ao 2 sh en m ie # 3 p ai z 2 l ei sh e 3 ch 2 # y # # sh e 9	6 33	6 31	58.8 319.7
贵吗 你们回吧 不复杂我都记住啦	g 9 # g ui m ar 4 4 # n # n im 4 h ui b u 3 # 4 b u 2 2 # f u z 2 w o 2 d ou 4 # d u # ou 2 j i zh u 2 1 3	11 13 28	9 11 25	130.3 139.2 301.1
告诉 36 存文档	g ao s 3 sh # 3 6 c un 2 w en d ang 4 5	17	16	215.3

and ease of use, which have been demonstrate in both healthy subjects and an ALS patient.

Acknowledgements National Key Research and Development Program of China (No. 2017YFB1002505), National Natural Science Foundation of China under Grant (No. 61431007), Key Research and Development Program of Guangdong Province (No. 2018B030339001).

References

1. J.R. Wolpaw, N. Birbaumer, D.J. McFarland, G. Pfurtscheller, T.M. Vaughan, Brain–computer interfaces for communication and control. Clin. Neurophysiol. **113**(6), 767–791 (2002)
2. X. Gao, D. Xu, M. Cheng, S. Gao, A BCI-based environ- mental controller for the motion-disabled. IEEE Trans. Neural Syst. Rehabil. Eng. **11**(2), 137–140 (2003)
3. G.R. Müller-Putz, C. Pokorny, D.S. Klobassa, P. Horki, A single-switch bci based on passive and imagined movements: toward restoring communication in minimally conscious patients. Int. J. Neural Syst. **23**(02), 1250037 (2013)

4. G. Bin, X. Gao, Z. Yan, B. Hong, S. Gao, An online multi-channel SSVEP-based brain–computer interface using a canonical correlation analysis method. J. Neural Eng. **6**(4), 046002 (2009)
5. Y. Zhang, P. Xu, K. Cheng, D. Yao, Multivariate synchronization index for frequency recognition of ssvep-based brain–computer interface. J. Neurosci. Methods **221**, 32–40 (2014)
6. A. Paris, A. Vosoughi, G. Atia, Whitening 1/f-type noise in electroencephalogram signals for steady-state visual evoked potential brain-computer interfaces, in *2014 48th Asilomar Conference on Signals, Systems and Computers* (IEEE, New York, 2014), pp. 204–207
7. B.J. He, Scale-free brain activity: past, present, and future. Trends Cogn. Sci. **18**(9), 480–487 (2014)
8. O. Friman, I. Volosyak, A. Graser, Multiple channel detection of steady- state visual evoked potentials for brain-computer interfaces. IEEE Trans. Biomed. Eng. **54**(4), 742–750 (2007)
9. E.W. Sellers, E. Donchin, A p 300-based brain–computer interface: initial tests by ALS patients. Clin. Neurophysiol. **117**(3), 538–548 (2006)
10. S. Silvoni, C. Volpato, M. Cavinato, M. Marchetti, K. Priftis, A. Merico, P. Tonin, K. Koutsikos, F. Beverina, F. Piccione, P300-based brain-computer interface communication: evaluation and follow-up in amyotrophic lateral sclerosis. Front. Neurosci. **3**, 1 (2009)
11. J.N. Mak, D.J. McFarland, T.M. Vaughan, L.M. McCane, P.Z. Tsui, D.J. Zeitlin, E.W. Sellers, J.R. Wolpaw, EEG correlates of p 300-based Brain-Computer Interface (BCI) performance in people with amyotrophic lateral sclerosis. J. Neural Eng. **9**(2), 026014 (2012)
12. M.J. Vansteensel, E.G.M. Pels, M.G. Bleichner, M.P. Branco, T. Denison, Z.V. Freudenburg, P. Gosselaar, S. Leinders, T.H. Ottens, M.A. Van Den Boom et al., Fully implanted brain–computer interface in a locked-in patient with als. New England J. Med. **375**(21), 2060–2066 (2016)
13. A.M. Norcia, L. Gregory Appelbaum, J.M. Ales, B.R. Cottereau, B. Rossion, The steady-state visual evoked potential in vision research: a review. J. Vis. **15**(6), 4–4 (2015)
14. Y. Inouye, R.-W. Liu, A system-theoretic foundation for blind equalization of a fir mimo channel system. IEEE Trans. Circuits Syst. I: Fundamental Theory Appl. **49**(4), 425–436 (2002)
15. T.A. Schonhoff, A.A. Giordano, *Detection and Estimation theory and Its Applications* (Prentice Hall, Upper Saddle River, 2006)
16. C.W. Baum, V.V. Veeravalli, A sequential procedure for multihypothesis testing. IEEE Trans. Inf. Theory **40**(6) (1994)
17. D.H. Brainard, The psychophysics toolbox. Spatial Vis. **10**(4), 433–436 (1997)
18. Y. Wang, X. Chen, X. Gao, S. Gao, A benchmark dataset for ssvep-based brain–computer interfaces. IEEE Trans. Neural Syst. Rehabil. Eng. **25**(10), 1746–1752 (2016)
19. X. Chen, Y. Wang, S. Gao, T.-P. Jung, X. Gao, Filter bank canonical correlation analysis for implementing a high-speed ssvep-based brain–computer interface. J. Neural Eng. **12**(4), 046008 (2015)

Highlights and Interviews with Winners

Christoph Guger, Brendan Z. Allison, and Kai Miller

Abstract The preceding nine chapters in this book presented an introduction and summaries of eight projects that were nominated for a BCI Research Award in 2018. In this chapter, we summarize the 2018 Awards Ceremony where we announced the three winning projects. We interviewed authors of these winning projects – Drs. Ajiboye, Tangermann, and Herff – and then wrote a conclusion with future directions, including BCI Hackathons and Cybathlons. We hope these chapters have been informative and helpful, and may have even helped to spark some new ideas.

1   The 2018 Winners

As with prior years, we announced the first, second, and third place winners as part of a major international BCI conference. The Awards ceremony for the 2018 BCI Research Award was part of the 2018 BCI Meeting in Asilomar, USA.

Hundreds of students, doctors, professors, and other people attended the awards ceremony to see who would win. The outdoor weather was excellent and everyone was in a good mood. The organizer and emcee, Drs. Guger and Allison, invited a representative from each of the nominated groups to join them on the stage. All nominees received a certificate and other prizes, and remained onstage as the winners were announced. Figures 1, 2 and 3 show some of the first, second, and third place winners accepting their awards as the awards were announced at the ceremony. The 2018 BCI Research Award were:

C. Guger (✉)
g.tec medical engineering GmbH, Schiedlberg, Austria
e-mail: guger@gtec.at

B. Z. Allison
Department of Cognitive Science, University of California at San Diego, La Jolla 92093, USA
e-mail: ballison@ucsd.edu

K. Miller
Mayo Clinic, Rochester, MN, USA
e-mail: kjmiller@gmail.com

© The Author(s), under exclusive license to Springer Nature Switzerland AG 2020   107
C. Guger et al. (eds.), *Brain–Computer Interface Research*,
SpringerBriefs in Electrical and Computer Engineering,
https://doi.org/10.1007/978-3-030-49583-1_10

First Place:

Abidemi Bolu Ajiboye[1,2,6], Francis R. Willett[1,2,6], Daniel R. Young[1,2,6], William D. Memberg[1,2,6], Brian A. Murphy[1,2,6], Jonathan P. Miller[2,4,6], Benjamin L. Walter[2,3,6], Jennifer A. Sweet[2,4,6], Harry A. Hoyen[5,6], Michael W. Keith[5,6], Paul Hunter Peckham[1,2,6], John D. Simeral[7,8,9,10], John P. Donoghue[8,9,12], Leigh R. Hochberg[7,8,9,10,11], Robert F. Kirsch[1,2,4,6]

Restoring Functional Reach-to-Grasp in a Person with Chronic Tetraplegia using Implanted Functional Electrical Stimulation and Intracortical Brain-Computer Interfaces

1   Department of Biomedical Engineering, Case Western Reserve University, Cleveland, Ohio, USA.
2   Louis Stokes Cleveland Department of Veterans Affairs Medical Center, FES Center of Excellence, Rehab. R&D Service, Cleveland, Ohio, USA.
3   Department of Neurology, University Hospitals Case Medical Center, Cleveland, Ohio, USA.
4   Department of Neurological Surgery, University Hospitals Cleveland Medical Center, Cleveland, Ohio, USA.
5   Department of Orthopaedics, MetroHealth Medical Center, Cleveland, Ohio, USA.
6   School of Medicine, Case Western Reserve University, Cleveland, Ohio, USA.
7   School of Engineering, Brown University, Providence, Rhode Island, USA.
8   Center for Neurorestoration and Neurotechnology, Rehabilitation R&D Service, Department of Veterans Affairs Medical Center, Providence, Rhode Island, USA.
9   Brown Institute for Brain Science, Brown University, Providence, Rhode Island, USA.
10 Department of Neurology, Massachusetts General Hospital, Boston, Massachusetts, USA.
11 Department of Neurology, Harvard Medical School, Boston, Massachusetts, USA.
12 Department of Neuroscience, Brown University, Providence, Rhode Island, USA.

Fig. 1 Brendan Allison, Christoph Guger, Kai Miller, Abidemi Bolu Ajiboye (winner of the BCI Award 2018), Leigh Hochberg, Sharlene Flesher, and Vivek Prabhakaran during the Gala Awards Ceremony

Second Place:

Michael Tangermann[1,3], David Hübner[1,3], Simone Denzer, Atieh Bamdadian[4], Sarah Schwarzkopf[2,3], Mariacristina Musso[2,3]

A BCI-Based Language Training for Patients with Chronic Aphasia

1 Brain State Decoding Lab, Dept. Computer Science, Albert-Ludwigs-Universität Freiburg, Germany.
2 Department of Neurology, University Medical Center Freiburg, Germany.
3 Cluster of Excellence BrainLinks-BrainTools, Albert-Ludwigs-Universität Freiburg, Germany.
4 Inovigate, Aeschenvorstadt 55, 4051 Basel, Switzerland.

Fig. 2 The jury with David Hübner and Michael Tangermann

Third Place:

Christian Herff[1], Lorenz Diener[1], Emily Mugler[3], Marc Slutzky[3], Dean Krusienski[2], Tanja Schultz[1]

Brain-To-Speech: Direct Synthesis of Speech from Intracranial Brain Activity Associated with Speech Production

1. Cognitive Systems Lab, University of Bremen, Germany.
2. ASPEN Lab, Old Dominion University, Norfolk, USA.
3. Departments of Neurology, Physiology, and Physical Medicine & Rehabilitation, Northwestern University, Chicago, USA.

Fig. 3 Christian Herff accepts his prize from the jury

These winning projects all presented BCIs for patients. They were validated in real-world settings and could lead to new ways to help people with aphasia or difficulty moving. The three winners this year came exclusively from the US and EU. This is slightly unusual, since most winning projects across prior years have involved people from Canada and/or Asia (mostly China and Japan). Two of this year's three winning projects involved implanted BCIs, consistent with trends in recent BCI research and earlier BCI Research Awards.

Dr. Guger concluded the ceremony by thanking the 2018 jury:

Kai Miller (chair of the jury 2018),
Natalie Mrachacz-Kersting (Winner 2017),
Vivek Prabhakaran,
Yijun Wang,
Milena Korostenskaja,
Sharlene Flesher.

2   Interview with Dr. Ajiboye

A. Bolu Ajiboye and Robert F. Kirsch of Case Western Reserve University, USA, in collaboration with Leigh R. Hochberg of Harvard Medical School and Brown University, USA, were on the team that won the top prize in the BCI Research Award 2018. Their winning project was titled "Restoring Functional Reach-to-Grasp in a Person with Chronic Tetraplegia using Implanted Functional Electrical Stimulation and Intracortical Brain-Computer Interfaces". We had the chance to talk with the

1st Place Winner of the BCI Award 2018: Restoring Functional Reach-to-Grasp using Implanted FES and Intracortical BCIs

Fig. 4 Dr. A. Bolu Ajiboye

winner, Bolu, about this project. It was the first study to combine an implanted human BCI system with implanted FES to restore both reaching and grasping in a person who had lost all functionality.

Christoph: Can you tell me a little bit more about yourself and your background?

Bolu: "Yes. My name is Bolu Ajiboye. I am an assistant professor of biomedical engineering at Case Western Reserve University in Cleveland, Ohio. I have a Ph.D. from Northwestern University in biomedical engineering and I have been a faculty member of Case Western for about 6 years. I also work for the Louis Stokes Cleveland VA Medical Center, and most of my work is related to brain computer interfaces (BCI) applied to persons with spinal cord injury to restore movement after chronic paralysis."

Christoph: You submitted your recent work to the BCI Award 2018. Can you tell me a bit more about the submission and what it is all about?

Bolu: "Yes. For the past thirty years, Dr. Kirsch and other colleagues have been developing a technology called functional electrical stimulation (FES). Essentially, this technology uses electrical stimulation to reanimate paralyzed muscles, so persons with paralysis can regain motor function, and can move again. What we have done recently as part of the BrainGate2 pilot clinical trial, which is headed by Dr. Leigh Hochberg at Massachusetts General Hospital, is we have been able to combine a BCI with this functional electrical stimulation technology. Now, what the BCI does, is record brain activity that is related to reaching and grasping. We then take this brain activity and we decode it or decipher it, to try to predict what movement the person with the paralysis wants to make. And once we are able to use our BCI to extract the intended movement command, we can send that movement command to the FES system.

So, we are essentially going around or circumventing the spinal injury. Typically, with movements, the brain sends a movement command, which goes through the spinal cord. But with the spinal cord injury, that movement command can't get to the peripheral muscles and nerves. So with the BCI and FES we have now gone around the spinal cord, such that a person with paralysis can now think about moving, and then the arm and hand will move in the same way that they are thinking."

Christoph: And your BCI Award submission showed how this approach worked with a patient?

Bolu: "Yes, in the submission we basically showed a proof of concept that we could restore brain controlled motor function or brain controlled movement to a single person with chronic tetraplegia who had paralysis of the arms and hands. This is the first person in the world who has been able to use the BCI and the implanted FES to restore reaching and grasping. As a result of our work, this person, who was paralyzed 10 years before joining our study, was now able to perform activities that we take for granted—such as drinking a cup of coffee or feeding himself from a bowl of mashed potatoes, or even reaching and scratching his own face."

Christoph: We just held the BCI Award 2018 Ceremony yesterday, and you won the first place. How do you feel?

Bolu: "We did submit our project to the BCI Award 2018 and we were fortunate to win first place. We were very grateful and very thankful that the committee saw our work fit to win first place. The party was great—you know, there was music and food. We produced the video that was played there and everybody got to see the work that we did. Everybody got to see our patient using the system. It was a great time, and we are very thankful to g.tec and the BCI Society, who sponsored the BCI Award."

Christoph: Do you think the BCI Award is important in the field of BCI, or does it somehow acknowledge your work more internationally or make it more visible?

Bolu: "I think the BCI Award is important in our field. Every year, it highlights some of the top research, the top work that is going on in our field, and it is important to highlight that for the general public and also for researchers within our field. We are obviously very ecstatic to have won it, and it is a very important thing for our group. We believe that really it is our participants who have made this possible, the people who are benefiting from our work, and so we really accepted this award on behalf of them."

3   Interview with Dr. Tangermann

Aphasia refers to an impairment of language abilities, and is usually caused by a stroke in the left hemisphere. About 20% of all first stroke patients remain with a persistent, chronic communicative impairment which has a large impact on their quality of life. Motivated by the success of recent BCI-supported hand motor training protocols, Michael Tangermann and his colleagues from the Albert-Ludwigs-University Freiburg, Germany, have implemented and validated a BCI-based closed-loop language training protocol for chronic aphasia patients after stroke. Michael Tangermann is the head of the jury for the BCI Award 2019.

Christoph: Michael, what's your affiliation?

Michael: "I am from the University of Freiburg in the south of Germany, specifically the Computer Science Department. My Brain State Decoding Lab is embedded into the Cluster of Excellence "BrainLinks-BrainTools." It not only provides funding for neurotechnology research, for which we are very grateful, but it also allows us to team up with other groups of the Cluster for truly interdisciplinary projects. The award-winning project, for example, is a joint endeavor with Mariacristina Musso and colleagues from the Neurology and Neurophysiology Dept. of the University Medical Center Freiburg."

2nd Place Winner of the BCI Award 2018: BCI-Based Language Training for Patients with Chronic Aphasia

Fig. 5   Dr. Michael Tangermann

Christoph: You were nominated for the BCI Award 2018 and won second place. Can you tell me a little more about your submission?

Michael: "A special aspect of this work is that we go into a new application field with BCIs: the rehabilitation of language impairments after stroke. This foray of my group has been inspired by the growing number of novel BCI-supported motor rehabilitation approaches, which recently have started to generate scientific evidence of their efficacy. By transducing some of the underlying ideas into the language domain, we were able to address language deficits (so called aphasia) instead of hand motor deficits. For this transduction, it was extremely helpful that I had been conducting research on auditory BCI systems for several years already."

Christoph: What are the results of your work?

Michael: "We have developed and tested a novel rehabilitation training protocol for patients with aphasia. Making use of BCI neurotechnology, we now are able to monitor in quasi-realtime some of the cognitive processes relevant in language tasks. This allows us to help a patient develop his/her language ability during the course of our training. Practically, our protocol foresees a patient training effectively for 30 h. The training is intensive, and the patient visits the BCI lab at least 15 times within a few weeks."

Christoph: How can someone imagine the language task? How does it work?

Michael: "During training, the patient is seated inside a ring of loudspeakers while we record his/her brain activity. After indicating a so-called target word to the patient, we play a sequence of different words using the loudspeakers. The patient's task now is to carefully listen to the sequence and detect the target words played amidst numerous other non-target words, which should be ignored. Each word stimulus elicits a transient response in the EEG signal. Our assumption is, that target words (if attended and processed by the patient) should generally elicit a different response than all the non-target words. If this is the case, and if we should be able to detect these differences, then we can provide positive feedback to the patient. Wrapping up, we hypothesize that this feedback reinforces a good language processing strategy."

Christoph: So, what do you see in the brain when the patient is focusing on a specific word?

Michael: "Using machine learning methods, we can analyze each and every of the so-called event-related potential (ERP) responses of the EEG that is elicited by a word played. Our algorithm is trained on the individual EEG signals of the patient. This training enables us to detect even small differences between target and non-target word responses in quasi-realtime, even though these signals are hidden within the strong noise, which is typical for any EEG recording. The patient, however, can support this effort by concentrating hard on the target words and trying to ignore all the non-target words."

Christoph: And what goals do you want to achieve in the future with your work?

Michael: "With the data we have collected so far, we are convinced that our novel BCI-supported language training can be beneficial for chronic stroke patients, whose stroke happened at least 6 months prior to our training. But of course we want to learn more about this and other subgroups of patients, e.g. those who show great improvement or who benefit less, and we want to understand why this is the case. A first step in this direction is to investigate which systematic changes in brain activity are triggered by our training. When I explain our novel paradigm, it sometimes surprises people to learn that our patients are not required to actually speak during our training, but only need to listen. At a second glance, however, many sub-processes relevant for language are involved both in processing and in production. Thus, we currently think very positively of this characteristic of our protocol, as it may allow for the participation of patients, who are not even able to speak any more due to severe strokes."

"Digging deeper, we are currently recruiting chronic aphasic patients from Freiburg and the surrounding area to prepare for a randomized study. That will allow us to understand the exact causes for language improvements that we can observe. While there exists the possibility that the improvements may simply be the result of the circumstances of the training process (like in placebo treatment), we hope that the brain signal-based, task-specific reinforcement is what actually triggers the change. Otherwise we would of course not need a BCI system."

Fig. 6 This portion of Dr. Herff's poster at the BCI Meeting 2018 shows that, when this event began, he was a nominee and not yet one of the winners

Christoph: Are you working with patients in the lab or in the field?

Michael: "At the moment, we are running all of our experiments either in the labs of my research group or in our partner's clinic in Freiburg. This allows us to control the procedures and the processes well, which is important at the current state of our research. On the long run, however, we would like to collaborate with rehabilitation clinics and speech therapists to run the training protocol with simplified setups outside our own labs."

4   Interview with Christian Herff

To investigate and decode articulatory speech production, Christian Herff and his team used Electrocorticography (ECoG) and presented the first direct synthesis of comprehensible acoustics only from areas involved in speech production. This is especially important, because these areas are likely to display similar activity patterns during attempted speech production, as would occur for locked-in patients. In summary, Brain-To-Speech could provide a voice and a natural means of conversation for paralyzed patients. We had the chance to talk with Christian about his BCI Award 2018 submission at the BCI Meeting 2018 in Asilomar.

Christoph: Christian, what's your background and where do you come from? Could you also tell me a bit more about your BCI Award 2018 submission?

Christian: "I am a computer scientist and I am currently working at the University of Bremen, but I am going to switch to the University of Maastricht very soon. I work in Brain-Computer Interfaces based on speech processes. We have been focusing on automatic speech recognition from brain signals for a while. But we present

3rd Place Winner of the BCI Award 2018: Brain-To-Speech

Fig. 7   Dr. Christian Herff

something very new in this project. This time, instead of writing down what a person was saying, we try to synthesize the speech directly from the brain signals. So, from the measured brain signals, we directly output speech as an audio waveform."

Christoph: And did you manage it?

Christian: "Yes, we did. I have some listening tests with me, so you could listen later if you like to experience firsthand how well it sounds right here. But, of course, we currently need implanted electrodes for that."

Christoph: Would you describe the implantation as the biggest challenge?

Christian: "Well, I mean, this is very lab environment, with many challenges. Prior work with speech is not continuous. It involves single words that people read out, so continuous speech is one of the bigger challenges. But, there are still a lot of other challenges to be solved. We were quite sure that we could reconstruct some aspects of speech, but we didn't even suspect that our results would actually be such high quality that you can understand the speech that we reconstruct from neural signals—that was quite flabbergasting to us, and we liked that a lot."

Christoph: How do you want your research to continue in the future?

Christian: "I think the most important step is to close the loop, so this is offline analysis. We recorded data and then went back to our lab and tried to do the synthesis, but if we can close the loop and have a patient directly synthesize the speech from their brain, that would be perfect. So that is our next step."

Christoph: Do you know Stephanie Martin from EPFL in Switzerland? She won the 2nd place of the BCI Award 2017 with her work "Decoding Inner Speech".

Christian: "Of course, I know her very well. I last met her in San Francisco, just two weeks ago. Her work is fantastic."

Christoph: You both work on decoding speech. How does your research differ from her research?

Christian: "Well, Stephanie discriminated between two words, but in imagined speech. In her earlier work, she was able to reconstruct spectral features of speech, which was an outstanding paper. We go one step further because, from those spectral features, we reconstruct an audio waveform from that. So that is taking what she has learned to the next step, I think."

Christoph: Which patients could benefit from this research?

Christian: "I think an implantation will be necessary for quite some time, so the condition of the patient would have to be quite severe. Because they would be willing to have an implantation, obviously locked-in patients would benefit. But I also think cerebral palsy patients who can't control their speech articulators but are of normal IQ could greatly profit from this. And I have talked to some of them and they are willing to get implantations if the device is good enough."

Fig. 8 Christian Herff presenting his BCI Award 2018 submission at the BCI Meeting 2018 in Asilomar

Christoph: Do you think implantation technologies could be realized in three years?

Christian: "Well, I do not see it in three years, but maybe in the next decade. I do not think this is too far off any more."

Christoph: You are presenting your BCI Award submission at the BCI Meeting 2018 in Asilomar here. What can I see on your poster?

Christian: "Sure, let me fast forward it so we can start from the front. So, what we did, we recorded activity while people were speaking. Then, for one new test phrase, we compared each bit of data to all the data we have in the training data, and then we picked the bit that was the most similar to the new bit and took the corresponding audio snippet. And we did that for each interval in our test data. So, for example, for the second bit, it is most similar to the "**a**" in "**Asian**" and for the next bit, the best match is the "**v**" in "**cave**". Then we used some signal processing to glue those snippets of audio together to reconstruct "**pave,**" which was the word a person was saying. So, it is actually a very simple pattern matching approach, but it gives us really nice audio."

5   Conclusion and Future Directions

In our discussion chapter from last year's book, we said that we were considering adding more interviews with BCI research groups. This year's discussion chapter featured interviews with the three winners. We wanted to provide readers with not just more details about their projects, but also their related work and associated experiences. For the first time, we decided to remove the analyses of trends reflected in the BCI Awards from the discussion chapter to present these interviews instead. We're welcome to feedback on this decision, and will probably include trend analysis in a future chapter or other publication.

Our discussion chapters have often featured some of our commentary on the field. This time, we'd like to present two activities that might interest readers: BCI Hackathons and Cybathlons. We understand that many readers are students and makers who'd like to participate, and more senior readers like faculty who may want to organize one.

Over the past few years, two of the editors (CG and BZA) have become increasingly active with BCI Hackathons [1, 2, 4, 8]. These BCI Hackathons are similar to conventional hackathons with software, where groups of students or other attendees form into groups and have 24 h or so to develop and complete a small project. BCI Hackathons are similar, but the organizers provide BCI systems and software for the students to use. At the end of the BCI Hackathon, a panel of judges reviews the BCI projects and selects winners. Some BCI Hackathons have had themes such as art, while others encourage general BCI applications. As our articles note, we are not at all the only people who organize hackathons. The articles present other groups such as NeuroTechX, a student organization that has hosted numerous hackathons. We've seen that BCI Hackathons are a great way to involve and engage students and get positive local publicity. If you're a student who might be interested, talk to one of your professors.

We also would like to encourage more BCI Cybathlons. These are activities in which disabled users compete using BCIs. Cybathlons, like BCI Hackathons, have become a common component of BCI conferences and are gaining attention elsewhere [3, 5–7]. Cybathlons can be more challenging to organize than BCI Hackathons due to the need to coordinate with patients who may have special needs. For example, persons who have difficulty moving—who are often the target users of BCI systems—often need someone to drive them to and from the Cybathlon. However, they are another way to engage students and the local public, and help show people how BCIs can benefit patients.

The rise of BCI Hackathons and Cybathlons are only two indicators of the growth of our field. BCI classes, publication, media publicity, conferences, workshops, and patient involvement have all increased as well [1]. We're delighted with the BCI Society's involvement with the BCI Research Award, and as of this writing, one author (BZA) is working with the BCI Society to develop awards for outstanding individuals within BCI research. These new awards should complement the BCI

Research Awards, which instead focus on projects, and thus provide new opportunities for the best minds in our field to earn recognition and inform the public about the latest advances.

As BCIs become more prominent, public engagement will become more important to attract the best new people into BCI research and present honest and positive information about BCIs. High-profile announcements of major BCI R&D projects from Facebook and Elon Musk over the last few years has increased attention to BCI research, but most people still don't use or think about BCIs, and know very little about what they can or cannot do. A single negative news story involving an unethical person or entity that misuses BCIs could spawn public fear and mistrust, thereby reducing research funding and interest in BCIs for different users. Positive public activities, as well as solid new research with professional media coverage, is essential to guide BCI R&D in the best directions and help these systems gain broader adoption.

Our BCI Research Awards and books have always meant to recognize and encourage the top projects in BCI research. One or more of the chapters may present a project in a direction that particularly interests you, or you may want to read all the chapters to get an overview of the field. You might find methods, ideas, or devices that you could adopt for your research or use in a class project. The chapters include a contact email for the corresponding author, and you could email the authors with questions. While we've enjoyed writing and editing these books—we also enjoy learning more about these BCI projects—there wouldn't be much point if nobody read them. We thank you, our readers, and look forward to seeing your BCI project someday.

References

1. B. Z. Allison, A. Kübler, J. Jin, 30+ Years of P300 BCIs. Psychophysiol. **57**(7), e13569 (2020)
2. C. Guger, B.Z. Allison, M. Walchshofer, S. Breinbauer, The BR4IN. IO Hackathons, in *Brain Art* (Springer, Cham, 2019), pp. 447–473
3. F. Lotte, M. Clerc, A. Appriou, A. Audino, C. Benaroch, P. Giacalone, C. Jeunet, J. Mladenović, T. Monseigne, T. Papadopoulo, L. Pillette, Inria research & development for the Cybathlon BCI series, in *International Graz Brain-Computer Interface conference*, 2019, September
4. M. Ortiz, E. Iáñez, C. Guger, J.M. Azorín, The art, science, and engineering of BCI Hackathons, in *Mobile Brain-Body Imaging and the Neuroscience of Art, Innovation and Creativity* (Springer, Cham 2019), pp. 147–155
5. S. Perdikis, L. Tonin, J.D.R. Millan, Brain racers. IEEE Spectr. **54**(9), 44–51 (2017)
6. S. Perdikis, L. Tonin, S. Saeedi, C. Schneider, J.D.R. Millán, The Cybathlon BCI race: Successful longitudinal mutual learning with two tetraplegic users. PLoS Biol. **16**(5), e2003787 (2018)
7. A. Schwarz, D. Steyrl, G.R. Müller-Putz, Brain-computer interface adaptation for an end user to compete in the Cybathlon, in *2016 IEEE International Conference on Systems, Man, and Cybernetics (SMC)* (IEEE, New York, 2016, October), pp. 001803–001808
8. A. Valjamae, L. Evers, B.Z. Allison, J. Ongering, A. Riccio, I. Igardi, D. Lamas, The Brain-Hack project: Exploring art-BCI hackathons, in *Proceedings of the 2017 ACM Workshop on an Application-oriented Approach to BCI out of the laboratory*, 2017, March, pp. 21–24

Printed in the United States
By Bookmasters